What others are saying about

MILLIONAIRE FENG SHUI

Discover the Hidden Science of Super Rich Business

Alisse Bradley has one of the most extraordinary minds I've encountered. Her unique education and research and development in the areas of wealth creation, quantum physics and the power of thought are like none other. Her unique wisdom and gift for communicating in a way you can understand make Alisse a powerful teacher and leader in her field. If you want to understand the secret to a successful business, this book is your solution.

Julie Lewin / Author of 'AreekeerA™ Vibration: Healing Yourself from Within' / www.areekeera.com

I always thought Feng Shui was just about the importance of closing the toilet seat and how you arrange your furniture. Alisse Bradley has opened my eyes to how powerful and complex it actually is. I discovered a step-by-step system to strategically build on the way I have intuitively approached business to have an even greater impact. Thank you Alisse.

Janet Beckers / Award winning Mentor, Speaker and Best-selling Author of 'Romance Your Tribe Online' / www.romanceyourtribe.com

Alisse Bradley's book, Millionaire Feng Shui, *is very timely. I use Feng Shui in my home and office as a matter of course. Reading Alisse's book gave me new insights, new ideas and new tools for increasing the power of my business through Feng Shui. What's great is that it doesn't matter if you 'believe' in it or not... because energy flow is energy flow. For example, if you have something keeping your office door from fully opening, this is blocking energy fully coming into your business. Straightforward. Clear. Useful. I invite you to take the time and clean up your life, your office and your business with this book as your guide. You won't regret it!*

Baeth Davis / Award winning Information Marketer, Palm Pilot for the Soul of Your Business™ and Intuitive Business Catalyst / www.yourpurpose.com

While reading your book Alisse, it was both captivating and inspiring. It felt like a fresh view on Universal principles and practices I could relate to, and I often found myself making a checklist of things to implement straight away. Your wisdom has shone through – I understood immediately what you were saying here in your book and I could not stop to put it down. It is not only a great reference book to keep returning to, but a map to a way of living a purposeful and balanced life, where abundance, beauty and inspiration come forth. It will be one of those books that always has a place on my bookshelf, that I will never give away – but I will always give copies to those who matter to me.

Luanne Mareen / Author, Speaker, Global Goddess Gatherer and Divine Business Mentor / www.goddessonpurpose.com

What Alisse shares in this book is a unique mix of ancient wisdom combined with proven modern wealth creation strategies. Her strategies are simple to understand and her step-by-step instructions are easy to implement. If you want to awaken the energy flow around your business and increase your personal success, then I highly recommend you read Millionaire Feng Shui.

Lyndsey Baigent / Keynote Speaker, Coach and Mentor at Party Plan Revolution / www.lyndseybaigent.com

Alisse has integrated her East and West with simple exercises and empowering stories, to clear underlying blocks to your flow of money. I love that she tells women it is ok to be abundant and successful. More and more we are looking to solve our own problems, get to the underlying issues, take personal responsibility and look at how to be clear on our purpose. Honouring your business is so important, using the core principles of Feng Shui - to be clear in your life and your surroundings - to see patterns clearly - to have skills written in a way that are clearly explained and easy to implement. Alisse as a person - you amaze me - for your sheer determination to find strategies to handle the blocks in your own life and I love that you have applied that drive and determination to your work with Feng Shui.

Marguerite Louis Amor / Speaker, Facilitator, Kinesiologist and Family Constellations Therapist / www.creatingyou.com.au

Alisse's wealth of knowledge and experience absolutely shines through in her work. The depth of information she provides and the ease with which she transmits this into actionable steps for the layperson is a rare gift. I have thoroughly enjoyed reading Millionaire Feng Shui *and using the hidden science Alisse reveals to increase my own business success. If you are interested in making positive changes in your life and work, I would highly recommend starting with this book.*

Assisi Chant / Word Whisperer and Photographer, Energy Empath and Resonant Resource Builder / www.resourceswithresonance.com

Alisse operates from her heart in the most authentic and passionate way. Her knowledge of Feng Shui is stunning, as is the manner in which she lives and breathes this way of life. There is something truly special about Alisse, which you'll discover here on the pages of this life-affirming book.

Sue Papadoulis / Award winning Journalist, Author and Founder of Publicity for Profit / www.publicityforprofit.com.au

What others are saying about

ALISSE BRADLEY

Speaker, Author, Business Coach and Designer

Working with Alisse helped me 'unleash my power' that was lying dormant within me for years. Prior to working with her, I was confused as to my direction in life and lacked self-confidence. I was also unclear on what my destiny was, though after working with Alisse, she was able to bring forth my 'inner' desires that I didn't even know existed, which in turn gave me both the courage and stamina to move forward, and to bring these goals into fruition. It is an absolute pleasure working with Alisse; she has a heart of gold and is able to build a quick connection with all her clients, allowing for an amazing friendship to develop. I recommend Alisse to anyone wanting to find clarity in their life and for people wanting to take their life or career to the next level.

Katrina Kavvalos / TV Talk Show Host, #1 Best Selling Author / www.katrinakavvalos.com

Alisse has provided for me wonderful guidance on finding the right path to wealth in line with my passion and values. I always thought that I couldn't be a successful businesswoman because I didn't have the same gutsy attitude that others had. However, having spent quite a few sessions chatting to Alisse, she has helped me define who I am and the unique gift that I can share in a way that still gets me noticed and brings clients to my business – the kind of clients that I want to work with and who want to work with me. Not only have my personal and business finances improved, the relationship with my children has also improved. Thank you Alisse, you have been a wonderful friend and coach.

Karen Thomson / Business Development Management / www.pulseecommerce.com

After meeting Alisse at a Women's Seminar, we employed her services to completely rebrand our business. We have an online business with a natural organic skincare and skin clinic on the Gold Coast. We have not only increased our business over the past two months, but furthermore the feedback we are receiving from our customers is super positive. I highly recommend coaching with Alisse.

Debbie Majella Nolan / Founder of Door of Your Skincare / www.doorofyouth.com

Thanks to Alisse I am finally shifting the areas of my business and life I have been stagnant in. She has shown me how to keep the "FAITH" and "TO NEVER GIVE UP." This experience has not only enriched my life; it has also given me more understanding of how to achieve success in all areas of my life. Thank you Alisse, you are truly on your path of greatness. You are brilliant and I am so grateful of my time with you.

Sandra Campbell / Handcrafted Jewelry Designer / www.facebook.com/edenayadesigns

Before I met Alisse, I was floating along doing stuff but not moving forward. Now I am finally really happy with my branding and business; I have been to the United States twice and Canada for the first time to further my training and passion, have become recognized amongst my peers and network team as a leader, and facilitated some major expos in Queensland and New South Wales. I have the confidence to know that whatever I believe I can do, I will most surely achieve it. Alisse is a person who lives what she teaches. I give her my highest recommendation if you wish to bring balance, peace, and prosperity in to your life.

Barb Free / Founder of Live in Synergy / www.liveinsynergy.com.au

Alisse, whatever you are doing it is just amazing! When I turned my phone back on after our session I had a call from a fellow coach. She signed up with me immediately on my Big Coaching Pack. Synchronicity?? I am starting to think there is more to this. Not to mention I have been feeling amazing and have so much energy. I am even looking forward to exercising!! Thanx so much!! Love & Peace.

Carolyn Greer / Raw Food Nutritionist / rawvival@gmail.com

Alisse, you have reignited my confidence to move forward in a way that truly represents who I am, a confidence to trust my inner self, and a confidence that transcends into the lives of others. My sense of self is stronger than ever, and I am once again feeling happy. I am very excited for what lies ahead for me, my family, and my business. Thank you from the bottom of my heart.

Karen Mudie / Holistic Life Coach / www.karenmudie.com.au

I have been fortunate to have met Alisse about a year ago and continued to do her courses. I have found that Alisse gives much more than is usually expected. Packed with value from the time spent in her courses, the content, and her continuing support. Her great knowledge in a range of different areas and her intuition to just know how to help people make the changes they so want in their lives. I have finally made so much progress it is amazing, after going around in circles no matter how hard I have tried to move on for years. Now finally I have moved through in business, my relationships, and I have become who I am today. I am now in a place where dreams can come true! Thank you Alisse; I am so grateful for all of your support and guidance! It is exciting doing my own change. And fascinating watching those around me also moving through and finally making amazing progress! It has been an amazing time of learning! Thank you, Alisse!

Wendy Skene / Owner Operator at Fairy Dell Shop / www. facebook.com/fairydellbywendy

When working with Alisse, she makes sure that her strategies and principles are explained in plain English, and this demystifies the entire process so that you intuitively start to see for yourself what works and why.

Wendy Chamberlain / Buyer & Vendor Advocate / www.followwendy.com

Working with Alisse has been one of the greatest drivers in boosting my career success over the past four years. She has a wealth of knowledge to share with entrepreneurs and business owners who are ready to start making real money. If Alisse gives you a piece of advice, you can "take it to the bank!"

Daisy McCarty / Brand Messaging Specialist and B2B Copywriter / www.daisymccarty.com

One of the greatest gifts Alisse has is listening to you; what you want, need, and desire then using Feng Shui to create those opportunities for you. The most prevalent lessons I learned from doing Feng Shui with Alisse was to stop blocking my path. I'd leave boxes near doorways or chairs in the walkway area, creating unnecessary obstacles to where I wanted to go. It is simple logical Feng Shui lessons like this that allowed me to open my mind and human being-ness to creating less stress and allow myself more, and so give myself permission to succeed in whatever I choose to do.

Katey Shaw / Entrepreneurial Development Consultant

ALISSE BRADLEY

Millionaire FENG SHUI

DISCOVER THE HIDDEN SCIENCE OF SUPER RICH BUSINESS

Because Success is an Energy Game!

FOREWORD BY MARIE DIAMOND

www.alissebradley.com

First published in Toowoomba, Queensland, Australia
2016 by Alisse Bradley International

ISBN 978-0-9953560-0-9
DEWEY NO: 650.12

Book Production Services: www.bevryanpublishing.com

Disclaimer
All the information, techniques, skills and concepts contained within this publication are of the nature of general comment only, and are not in any way recommended as individual advice. The intent is to offer a variety of information to provide a wider range of choices now and in the future, recognizing that we all have widely diverse circumstances and viewpoints. Should any reader choose to make use of the information contained herein, this is their decision, and the author and publisher do not assume any responsibilities whatsoever under any conditions or circumstances. The author and publisher do not take responsibility for the business, financial, personal or other success, results or fulfillment upon the readers' decision to use this information. It is recommended that the reader obtain their own independent advice.

The client examples provided in this book represent actual clients and the outcomes they have experienced, however their details, including personal and business names and types have been changed for reasons of privacy.

This book is dedicated to my incredible husband, Paul, who always challenges me to become a better person, and to my four amazing children, Anthony, Meagan, Renée, and Caleb, who are my inspiration for living out my dreams.

I also dedicate this book to my beautiful friends: Julie, who is my 'Earth Angel'; Danette, who encouraged me from the beginning; Daisy, who empowered me at my every word; and Sally, who supported me to turn this book into a reality.

FOREWORD BY MARIE DIAMOND

Marie Diamond is known worldwide for her leadership in personal and business transformation. She is a respected author, consultant, teacher, and speaker with expertise and knowledge in diverse areas from Feng Shui energy to Quantum Physics and the Law of Attraction. Marie's current vision is to enlighten more than 500 million people around the globe through her powerful energy systems.

I met Alisse Bradley in 2009, when I chose to attract eight individuals to personally mentor. Through this relationship, Alisse has come into even greater connection with her heart and is realizing her own true value as a leader. It has been a great joy to see her building the success of which she has always dreamt.

Alisse Bradley is one of the powerful women on this planet who is here to enlighten the world with her wisdom, and soften problems with her Heart and Light. As her mentor and friend, I have seen how she generously shares her insights with the world.

This book will bring you to the next level of Greatness. Her tips and information will give you hope and give you tools to reach your full potential.

Enlightened regards,

Marie Diamond, *Feng Shui Master – Enlightened Leader and Featured Teacher in* The Secret
www.mariediamond.com

A WORD FROM ALISSE ABOUT USING FENG SHUI APPROPRIATELY

There are various **Feng Shui systems** you can work with … I've enjoyed studying the works of (and studying alongside) many of the **world's masters**; however, I implement and teach **ONLY** what works **effectively for me**.

I do not **'sell out'** to Feng Shui as a **set of rules** I must follow. Instead, I experience it as:

- An elite opportunity to tap in to **'Ancient Wisdom'** that has proven to have **made sense** for thousands of years
- An **incredible awareness** of the **'Flow of Energy'** that forms our entire Universe
- A system of **'Self-Empowerment'** to **create my ideal reality** in every aspect of it

For me, it is a **success strategy** I use in **my business** and **my life**: I hope it is the same for you.

TABLE OF CONTENTS

PART THREE

INTRODUCTION

The practice of Feng Shui dates back over 4,000 years in Eastern traditions. It has cultural, philosophical, and mystical components. However, the main thing you will notice as you read this book is that Feng Shui is mostly common sense!

You don't have to subscribe to a specific set of 'religious' beliefs to make this practice work in your life. Feng Shui will resonate with you because, deep down, you already possess this complex spiritual wisdom. Now, you are simply finding an easier way to access it.

WHAT IS FENG SHUI?

At its most basic level, Feng Shui is about understanding and influencing the flow of energy (Chi) in your surroundings. Everything in our universe is energy in one form or another, so it isn't surprising that this ancient art and science can impact every area of your life.

The term *"Feng Shui"* translates in English as wind and water. Wind represents your inner (spiritual) environment. Like the movement of air, you can't see it, but you can feel it and see its effects in your life. Water characterizes your outer (material) environment. You can see it, you can touch it, and ***you can also direct its flow***!

How Does it Work?

These two elements of wind and water are intimately connected to one another. Any change in your inner self will impact your outer experience. If you've worked on improving your 'inner environment' using positive thinking or other self-development systems, you already have some idea of how this works.

What we sometimes forget is that the feedback system works in the other direction too, completing the cycle. Our outer reality (including our workspace, our belongings, even the clothes we choose to wear) affects our inner landscape. Even a small change can have a significant impact.

Intuitively, you already know this. Think about how you feel when your desk is cluttered:

- **Frustrated**
- **Overwhelmed**
- **Distracted**
- **Drained of energy**

Now, consider how you feel when everything is organized neatly in drawers and trays:

- **Calm and relaxed**
- **Balanced and in control**
- **Focused and energized**

Why is there such a big difference? All the 'stuff' that was cluttering your desk is still located within the same few cubic feet of physical space. However, you have created an intention: *"I have a clean, well organized workspace,"* and ***taken action*** to change your surroundings to reflect your intention. This effectively changes your inner experience of your outer environment.

Feng Shui provides a framework within which to explore and exercise your intuition. It increases your awareness of what your current surroundings, habits, and preferences reveal about your inner state of being. Then, it provides practical tools you can use to create the positive energy flow you really desire. This changes the vibrations (feelings) you experience internally.

Feng Shui and Quantum Mechanics Have a Lot in Common

The fact that consciousness and matter are inextricably intertwined, each affecting the other, has now become inescapable. Science is just starting to catch up with what each of us has known intuitively all along. As the field of quantum physics matures, we will see even greater alignment between newly 'discovered' scientific principles and ancient wisdom. You can expect an exponential increase in the knowledge and self-empowerment that is available to all of us for use in this new era of purposeful and conscious business.

One of the most important truths being confirmed with advances in the field of quantum theory is that reality is determined by what we **perceive** it to be (whether perception creates our reality or whether it simply confirms our experience of reality is largely immaterial in day-to-day living). That sounds obvious, but we often forget that we can only *be* as wealthy and prosperous as we *feel*. It is not possible to experience wealth in the future if you perceive nothing but impoverishment right now. This 'broke' mentality has to be overcome first. You must begin to live wealthy today, **in this moment**, rather than just wishing or working for a future time when you might finally achieve wealth. Bear in mind that wealth is not just money. True wealth represents freedom and security, the two things we require to express our authentic selves to the fullest. There's no time like the present to begin living authentically!

Feng Shui assists you in understanding how you currently perceive reality and prompts you to make concrete changes in your outer environment that trigger new perceptions. Being aware, purposeful, and connected in this way lets you harness your power as the conscious observer of reality to create what you desire.

WHY FENG SHUI FOR BUSINESS?

Chi Affects Everyone

Like ripples on a pond, the energy that surrounds you and flows from and through you also spreads out to connect with others. This includes your current and potential clients. With so many choices facing the average consumer, it's not surprising that people use their initial feeling or intuition to decide whom they will do business with. When you take ***conscious*** responsibility for the overall feeling, energy, or 'vibe' you are projecting, you will be amazed at the wonderful clientele and business opportunities this brings your way. We'll explore this idea more, later in the book.

Feng Shui is Serious Business

So, who actually uses Feng Shui for business and career success? A huge number of well-known entrepreneurs, millionaires, billionaires (including Bill Gates), global corporations, and celebrities from Oprah to Madonna use Feng Shui regularly. Here's just a short list of well-known businesses that rely on this art and science in differentiating themselves from the competition, increasing earnings, and creating long term business success:

- Chase Manhattan
- Citibank
- Coca-Cola
- FedEx Express

- Fujifilm
- Honeywell
- Hyatt Hotels
- Marks and Spencer
- Morgan Guaranty Trust
- Motorola
- Nike
- Paine Webber
- Procter & Gamble
- Shell
- Sony
- The Body Shop
- Virgin Airways

Seasoned corporate executives aren't turning to Feng Shui as part of a fad or wishful thinking. Donald Trump (who used Feng Shui in the design and construction of Trump Tower) is actually known for saying, " I don't have to believe in Feng Shui, but I use it because it makes me money."[1] Superstars in the business world have no time for using a system that doesn't work. They want proven results, and Feng Shui keeps coming through for them.

Ironically, if you scratch the surface of most successful entrepreneurs, you will find they are applying Feng Shui whether they recognize it or not. That's because Feng Shui is not just a set of rules. It's a system that is based on the underlying mechanics of the science of success. The natural laws, elemental cycles, and universal principles behind the ancient Eastern art of Feng Shui are what make it so relevant for Western business. As a rule, everything nature does is for the success of the individual within a greater system. It takes into account the environment, climate, competitors, etc., and arranges the individual with reference to all of those things so that it has the best chance of success. As a result, successful businesses make a habit of hiring the best in their industries, those individuals who have organically tapped all the fundamentals of what good Feng Shui is. This includes passion, purpose, planning, positivity, and productivity; everything that makes a business attractive and magnetic.

When you see how everything is connected, it all starts to make sense.

Get a Head Start

There are many methods to discover these truths and many entrepreneurs find their way intuitively by trial and error. The proven business science of Feng Shui you will learn in this book simply gives you shortcuts so you can get right down to building a successful business. You will be using the same professional Feng Shui strategies that have already been abundantly demonstrated as highly effective by modern, million dollar businesses. In the process, you will become your own guru by using your intuition to discover which techniques work best for your own unique business.

This resource is much more than an overview of Feng Shui principles and philosophy. You will get specific, practical information about what colors, symbols, shapes, positions, and other aspects of your environment really mean. You will learn how to use Feng Shui as a business awareness tool and explore topics ranging from finding your true purpose and passion to effective branding and design. You will also learn how to make your home office or business premises into the perfect work environment. Then, ***you*** can decide how to update the physical elements of your business (in conjunction with your intentions) to create and attract your deepest desires.

Furthermore, the processes you will be exposed to are pathways to discover the 'unconscious competence' embodied by superstar businessmen and women. By unpacking what they do so well on an **energetic level** there is the profound opportunity to recognize your own 'unconscious incompetence' and turn this into a 'conscious competence.' This means what comes so *easy* and *organically* to the successful becomes *learnable* and *available* to anyone prepared to embrace the text in this book.

To achieve a different result, we must ***do*** something different, but in order to do something different we must first ***know*** something different. This book is going to teach you *something different*.

PROLOGUE

THE NEW AGE OF BUSINESS

The face of the marketplace is changing very rapidly as more and more people are awakening to the fact that doing 'business as usual' just isn't worth the price of a lifetime of work. We crave meaning now more than ever. For many of us, business is no longer just a means to gain income. It is a journey we commit to. It is a path we choose that leads us to our deepest desires in life. There are challenges on the way that assist us in learning valuable lessons about ourselves and also about our ability to overcome obstacles. The reward is far more than any amount of money can buy. It is our fulfillment on a deep soul level.

Feminine Energy is Awakening

Women are at the forefront of this different way of doing business. In the U.S., Canada, and Australia, the number of women-owned businesses has been steadily increasing at about twice the average rate for other privately owned firms. This rush of innovative feminine energy into the field of business is bringing with it qualities of compassion, cooperation, and community. Because each person (whether male or female) has both a masculine and feminine energy within, this 'Age of Aquarius' is providing a perfect opportunity to bring these aspects of ourselves into a true state of balance. Entrepreneurs and small business owners are blessed with the ability to make this shift more easefully and quickly than big businesses. So, be prepared to lead the way!

The following will drive the *New Age* of Business:

- An understanding of Oneness. We are all one of the same *Source of Universal energy* and we are all connected.
- The knowledge that business can and needs to be Conscious, Compassionate and Cooperative; not Comparative and Competitive.
- An appreciation for how technology is making online business the wave of the future, allowing us to create powerful communities of like-minded business owners.
- The arrival of abundant new discoveries linking science and metaphysics.

Today's entrepreneurs and business owners are increasingly motivated by:

- Authenticity, good will, openness, honesty, and love.
- A desire to be of service to humanity.
- Love for the environment.
- Tolerance for all.
- An ability to let go of control and simply trust.
- Inspiration, innovation and individuality.
- An understanding of the true value of one's business.
- A shift from ego to spirit.
- An expectation that business must be fast, exciting, and rewarding.

To get on board with this shift, be prepared to:

- Say "Yes!" to change.
- Experience and express gratitude daily.
- Become self-aware.

- Operate from right intentions.
- Take responsibility for where you are right now.
- Connect with your higher self.
- Learn how to intentionally materialize what you desire.
- Let go of old fears and issues.
- Care for your physical and mental health.
- Pay attention to your dreams.
- Have the courage to step out and start new projects.
- Express your uniqueness.
- Access your infinite ability.

Note from Alisse: As with most people, my journey of self-discovery has been long and I expect it will continue for a lifetime! I have always been driven by a deep desire to progressively evolve, expand my mind, and overcome my life challenges. I've struggled with chronic health issues, low self-esteem, weight gain, being constantly broke, and sacrificing time with my family to try to make business endeavors work.

Through years of experimentation with various energy systems, I've found one thing holds true. My business can only grow as much as I grow. Once I connected with my purpose and passion it was only a matter of continuing to discover tools that worked to support my reason for being here. Feng Shui is one of the modalities that I have found to be enormously helpful in creating lasting business success in alignment with my deepest desires. I have used **Feng Shui** to overcome my challenges and assist me in:

- Knowing what I really want
- Being able to make clear and confident decisions
- Trusting my own judgment and abilities
- Knowing where to start and what to do next
- Valuing myself and my services
- Getting real results in my business

Today, I literally live the life of my dreams. I have my own business that is perfectly suited to my passion. I have achieved balance in my work and family life so I can run a successful business and still be present as a mother for my four precious children. I now make real money and have all the resources I need to make the choices I want. My husband and I both love our work and lives; we love that we can invest time into our own successes as well as spend time together taking holidays, laughing, and building wonderful memories with our family. I have found that the more I value myself, the more authentically I live. The more I serve humanity, the more abundance pours into my life. I believe this new way of doing business is available to everyone.

I wrote this book because I do not subscribe to the idea that success is based on following rules, regulations, or gurus. I rather embrace what I can learn from them, then I follow my own heart and soul to discover what success looks like for me. I love to support other business owners in realizing their own limitless potential in the same way.

May it be so for you!

GET READY TO CHANGE – FOR THE BETTER

As you read this book, bear in mind that ***everything*** is energy (Chi). You can apply what you learn here to all aspects of your life and start getting tangible results. Be prepared to experience:

- **Growth that will amaze and energize you**
- **Prosperity in every aspect of your business / life, inside and out**
- **Abundance that you can joyfully share with others**
- **Wealth that reflects your *true value* and self-worth**

- **Freedom to pursue your passions and express your authentic self**
- **Satisfaction beyond anything you've ever known**

Defining and visualizing what true success looks like is your first conscious step on the path to making your (inner) dreams an (outer) reality. The ability to achieve your every desire is already present within you. Throughout this book you will be given action steps to complete that bring you into alignment with your business. Here's one to start off with that will help open your mind and heart to your own infinite potential.

HOW YOU WANT YOUR BUSINESS AND LIFE TO BE

Since true business success connects with all areas of your life ...

Write down exactly 'How You Want Your Business and Life to Be!'

Keep in mind that this is your space to express exactly 'how **you** want it to be' without any limitations. To share all the joy and passion you can muster. You don't need to know 'how' you will achieve this right now; this exercise is for the purpose of expanding your imagination, so simply write down your deepest desires, your story.

Here are some life areas to guide you through this process:

- **Self / Spirit**
- **Career / Life Path / Legacy**
- **Money / Wealth**
- **Relationships / Family / Community**
- **Health / Wellness / Hobbies**

Self / Spirit:

Example: *My business allows me to do what I want when I want. I get up to my own alarm and write my own schedules. I now have the time to start every day in a way that nurtures me. I visit my purpose, vision, and goals for my business and my life, meditate, go for a morning walk in nature, and then share a nutritious breakfast with my husband and children.*

Now your turn:

Career / Life Path / Legacy:

Example: *My business has provided me a pathway to speak all over the world, write books, create products, and administer programs inclusive of my own research, philosophies, and experiences. I coach entrepreneurs and organizations how to build successful businesses and live fulfilled lives and I love that I empower others by helping them grow to all they can be, leading them on their own journey to realize their dreams.*

Now your turn:

Money / Wealth:

Example: *My business earns well in excess of 6 figures per annum and continues to grow every year. I now have ample money (x dollars) coming into my bank account each week. This has provided me with enough funds to buy a brand new car, fill my wardrobe with designer clothes, pursue the hobbies I love, travel and holiday to all the places I have dreamed of, continue to study and improve myself personally, expand my business and share it with a greater clientele, and make wise investments to take care of myself and my family for the rest of our lives.*

Now your turn:

Relationships / Family / Community:

Example: *Thanks to my business I always feel so happy and full of life. It rejuvenates and energizes me and this overflows into all of my family, friend, and community interactions. I have the time and energy to spend with each of my family members individually every week, to invest and enjoy time with my friends, and to contribute back to my community both financially and personally. I feel so blessed that I now have the inner and outer resources to give back to what matters most to me.*

Now your turn:

Health / Wellness / Hobbies:

Example: *Caring for my own health is of high priority in my business and my life. My business affords me the time and money to nurture myself with weekly massages, beauty treatments, hair designs, and relaxation therapies. I swim and sit in my sauna almost every day, as well as work out in my home gym. I have the time and money to buy and prepare nutritious foods and have employed a personal trainer to assist me with my daily exercise program. I always feel so strong and empowered.*

Now your turn:

____

____

____

____

____

____

____

____

I feel so blessed and deeply grateful that this is my life, all of this or ***something better****.*

KEEP THIS EXERCISE CLOSE AS YOU WORK THROUGH THE PAGES OF THIS BOOK.

PART ONE

Chapter 1
BECOMING HUMAN

Discovering Your Purpose Through Your Distinct *Vibrational Pattern*

Each one of us is a unique energetic being and we are all part of something much larger than ourselves. Every individual has something special to offer in this lifetime; this is your Life's Purpose. Your mission is the fulfillment of your Life's Purpose. This gift that you offer the world (in your life and in your business) is an outpouring of your authentic self, an expression of your true potential.

What This Feels Like

When you are in alignment with your Life's Purpose, you are open to an unlimited flow of energy. You may sense that a greater force is working through you (always in a positive way). This sense of connection and fulfillment will deepen as you continue to walk your innate path. You will feel that you are bringing a message of inspiration to everyone around you in everything you do.

Does this whole idea seem a little grandiose? It really isn't. When you are fulfilling your Life's Purpose, you will clearly see that every individual has this same potential. Every person has their own special gifts and the natural capacity to be truly remarkable.

WHAT ARE YOU HERE FOR?

Using Feng Shui to assist you in building a successful business is wonderful. However, no amount of money or wealth will bring you fulfillment and true abundance. This only comes when your work is in alignment with your Life's Purpose. It is never too late to begin living what you really came here to do. So, don't let past choices hold you back from changing your course in your career and your life to put you on the perfect path.

Although your specific Purpose will be as unique as your fingerprint or your DNA, every human on this planet has three things they are here to do:

Firstly, we are here to discover (or remember) who we are. Deepak Chopra would call this 'discovering your true self.'[2] Other spiritual authors might refer to this as living authentically. In Native American traditions it might be spoken of as *'becoming human.'*

You have probably met or seen people like this; they practically shine with an incredible, vibrant energy. Super successful business owners often fall in this category. They aren't 'perfect' people who never make mistakes, but they have realized they are perfect reflections of the Creative Source of the Universe just the way they are. When you acknowledge and accept yourself in the same way you are in the process of becoming who *you* truly are.

The second thing we are here to do (once we KNOW who we are) is to *express* who we are. Each of us has a distinct vibrational pattern, like a musical note, unlike that of anyone else. Like a beautiful symphony, we must all be present for the harmony of the Universe to be complete. Expressing your creative ability is like playing your musical instrument instead of leaving it to gather dust on the shelf. When you are expressing your talent, it is something that brings you pure bliss. This is not about what anyone else thinks about you or wants from you; this part of your Life's Purpose is all about you.

Being of service to others is the third thing you are here to do. This has nothing to do with subjugating your will and desires to anyone else. Instead, it is about acting in the full knowledge of **who you are** to use your **creative talents** in sharing joy with those around you. Giving and sharing in this way is not burdensome; it is one of the most delightful things you can do. When your business is centered on this model, your passion for what you do will never fade.

Client Story: Sarah is a client who has a beautiful home, a luxury car, and money to spare from her husband's successful business. She is a mother who has a good relationship with her husband and kids. She seems to 'have it all.' Yet she often finds herself sitting by her pool crying because she has a deep drive to share more of herself, and she isn't doing it. She feels guilty over this sadness. She thinks perhaps she isn't being grateful for what she has. But she has a deep purpose inside that she wants to fully express. Now, she has started her journey by writing a book which is finally giving her the chance to express herself.

Are you holding something back that you would really love to share with the world?

HOW CAN YOU DISCOVER YOUR SPECIFIC LIFE'S PURPOSE?

Remembering why you are here can be accomplished in lots of different ways. There are many de-coding systems that can help you clarify this purpose. Giftedness and Personality Profiling, Numerology, Planetary Birth Charts, Birth Order Systems, Chinese Astrology, Iris Diagnosis, and Hand Analysis are just some. You may decide to seek assistance from a specialist in one or more of these areas when remembering your Life's Purpose. This can be a powerful way to jumpstart the process of self-discovery.

Be sure to choose a professional who is obviously fulfilling their own Life's Purpose in providing this type of service. You could do this by asking them a few simple questions, like:

- What compelled you to get into this line of work?
- What do you love most about practicing this art / science?
- What do you most want your clients to take away from your services?

It won't take you long to get the general gist of whether this professional is operating from their purpose and passion or not.

Furthermore, you may tap into your own intuition and emotions to identify your authentic path. Whether you realize it or not the knowledge of your Life's Purpose (like all other wisdom you require) is already present within you. Your passions, your talents, and the activities that offer you the highest sense of fulfillment reveal your purpose. What makes you feel excited, inspired, and irrepressibly happy? What are you really good at doing that feels effortless and enjoyable? Those are important clues. You will find specific questions at the end of this chapter to help you explore these ideas further.

WHAT DOES PURPOSE IN BUSINESS FEEL LIKE?

In your business, living on purpose is all about moving (as Wayne Dyer referred to in his book *The Shift*[3]) from a motivation of Ambition to a motivation of Meaning. The vibration associated with 'scrambling to get ahead' is very different from the vibration that occurs when you are simply expressing your infinite ability.

Inspirational writer Earl Nightingale (*The Strangest Secret*[4]) spoke of success as "the progressive realization of a worthy ideal." Remember that you are trading every day of your life for something. Trade these days for something worthwhile and you will be getting a bargain!

Your purpose in life will always be something you can express in all areas of your life, including your business. After all, as an entrepreneur you are probably spending 40% or more of your waking hours on your business. It makes no sense to spend time doing this so you can do what you are *really* here for in your 'spare time.'

There is no such thing as spare time, there is only right now. Your **business** needs to support and nurture you in the following ways:

- **Developing your infinite potential**
- **Sparking your creativity**
- **Demonstrating your talents**
- **Stoking your passions**
- **Promoting inner peace, joy, and love**

Are You in the 'Wrong' Line of Work?

Client Story: Claire is a client who literally used to throw up every morning before she went to work. She was so out of line with her purpose that her body was telling her in no uncertain terms that she could not stomach her job any longer and needed to follow her passion. She now does and everything flows in her personal life and her work. She has a successful business that she loves.

Are you listening to the clues in your life about whether your business is nurturing you?

Do you find yourself meeting one obstacle after another in your business? Perhaps you feel frustrated as your every effort to grow and expand is thwarted. You have poured so much energy into your company that the very thought of 'giving up' and trying something else is too hard to contemplate.

However, as you seek out your Life's Purpose you will have to challenge your belief systems and decide if they are truly your own. This means that you may need to rethink your business model as well. That's OK. You are making progress toward becoming your authentic self.

This is the perfect time to look at your current career. Are there changes you could make in the nature of your product or service that will bring your business more in line with your true path? Perhaps you would resonate better with a different delivery method or target demographic.

Be willing to examine the possibility of completely new, radical choices. Continuing to do the same thing over and over out of habit or fear won't get you where you want to go. This doesn't mean you have to rush into anything. Take as much time as you need to mine your intuition for the perfect answers. Then, trust yourself to know what is best.

No Regrets

There is no such thing as failure in your life, only feedback. No matter what business ventures you have tried in the past, they represent valuable life lessons. Don't feel discouraged if you have spent years or even decades pursuing career goals that did not directly serve your ultimate Life's Purpose. That process provides you with an excellent contrast. It allows you to appreciate the feeling of bountiful, joyful, abundant living even more when you finally achieve it, and chances are you learned a lot of information and skills that will directly benefit you in your new endeavors.

Change Your Focus

Maybe you are already in the perfect line of work for fulfilling your Life's Purpose but don't realize it. Sometimes, all you need to do is change your perspective. You can accomplish this by rethinking what you are really offering your clients. No matter what product or service you provide, it serves much more than is first obvious. Let's look at a few examples:

Real Estate Agent

If you work in residential real estate your core purpose is not to move housing on and off the market. It isn't even primarily about making sure your clients have a place to live. You are actually facilitating individuals and families in creating their own ideal living environment. This will be one that speaks to their souls and brings them joy, comfort, and a sense of security. In short, you are helping them find a HOME that reflects who they truly are.

Personal Trainer

As a personal trainer your core purpose is not helping people lose weight or fix their diet. It also isn't really about increasing their strength, stamina, and flexibility. At the most basic level, you help human beings connect with and honor their physical self. You are helping them explore their body and remember that it is another manifestation of their unique energy. You are assisting your clients in ensuring their physical vibration reflects who they truly are.

Financial Advisor

This is another enormously valuable service. Your business is not solely about drawing up planning documents and advising on trust accounts. Your true service is to teach people the powerful energetic value of money and how to manage the flow of money in and out of their lives.

As a financial consultant you are in the unique position to assist clients in clarifying and healing their relationship with money (which is a very powerful form of energy). Your sound advice could very well be one of the keys that finally unlocks their prosperity potential.

Do You See a Pattern Emerging?

It is awe-inspiring to realize that at the heart of every worthwhile enterprise is the same type of wonderful, energetic purpose. How do you know that a particular business venture is right for you? When it is the one that raises your vibration and helps you express who YOU truly are while doing the same for others.

A Purpose Powered Business

Put your heart into following your dreams. When you connect with your purpose and decide to bring it into your work, you will discover there are no limits to your success. You will be motivated, passionate, and full of energy.

Don't create roadblocks in your path by limiting yourself based on your current position. When you make a firm decision about your business prosperity, the Universe will respond by supporting you in your intentions. Be prepared to find that **everything** is possible.

Client Story: Jean is a client that has been working in a service job for years. She is really good at what she does and very valuable to her company, but nothing flows. She just scrapes by with her income. She does not get to do the things she wants to do and she feels unhappy all the time.

When she realized her Life's Purpose it became obvious to her that she was not a failure. She was just bored with her life (not inspired or 'in-spirit'). She started considering the possibility that there was a business she had always wanted to establish in the healing field. She used Feng Shui to clean and beautify her home. She enrolled in a relevant course to pursue her new career even though she did not feel she had the money to do so. Suddenly, her finances started to flow much more easefully and supported her to make the change.

Are you open to considering completely new possibilities and directions for your career?

YOUR DEFINITE PURPOSE STATEMENT AND BUSINESS VISION

By now it's obvious that you need to find a way to define your Life's Purpose before moving forward. One way to focus and refine the concept of your Life's Purpose is to boil it down to what you can think of as your **Definite Purpose Statement.** This statement accepts that your ultimate purpose is to experience and express deep and abiding joy at the soul level (as opposed to temporary feelings of happiness). It acknowledges that you are here to develop your skills, talents, and abilities by doing and being what you love, and sharing it with others. You will then use this Definite Purpose Statement to develop a **Business Vision** that is in full alignment with your purpose. This vision articulates the *details* of how you make your expression of joy a service to others.

Wow, that sounds really different from the thinking behind the 'mission statements,' 'values statements,' or 'vision statements' of most standard companies today. A typical mission statement is crafted by a PR department, rather than being based on the actual purpose and values of a business and its owner. You can always tell the difference by looking at what a company does rather than what it says. As an entrepreneur, you are in a unique position to ensure your statement reflects what you truly value, your real mission, and your most exciting and extravagant vision. You can double check your daily activities and your business goals against your **Business Vision** to make sure you are staying on track. Your vision becomes a very powerful tool that energizes every area of your business and provides clear direction.

Client Story: Tina is a client who has ideas galore. She has this concept and that concept and five websites with different products and services. She gets frustrated because she wonders if she should be doing just one thing. She can't decide which idea to focus on. Once she identified her definite purpose, she realized that all her concepts revolved around the same central vision for her customers. She was able to tie everything together into one definite purpose and it made her direction and goals understandable and workable for her.

Would a single, clearly articulated Definite Purpose Statement finally help you to find your best direction?

The Process Overview

At the close of this chapter you are going to answer important questions that will help you find your strengths and passions, create your Definite Purpose Statement, and define your Business Vision. Here's a real life example (kindly shared with us) of what this process of self-discovery is like for one client. It begins by summing up all the information gathered about a client named Sarena.

Sarena's Definite Purpose Process

Sarena's Passions: Sarena **loves** working with aromatherapy and essential oils. She loves to research and learn; her books are very special to her. She loves the planning and designing elements in her business and in her home. She loves being of service and a part of the healing process for people. She loves spending time with her friends, listening to music, and watching movies as well as having space to relax and rejuvenate. She especially loves any time enjoying and working with nature, such as planting and gardening.

Sarena's Strengths and Talents are healing, therapeutic massage, working with aromatics and nurturing others, high level intuition, an ability to research and learn, an ability to create, color, and design. Sarena is great at writing, photography, singing, arts and crafts, and creating a beautiful home. She also excels at creating and producing natural healing products.

Sarena gained extra insight by having her **Planetary Birth Chart** drawn up by a professional. It confirmed that Sarena is a brilliant communicator, writer, networker, and student of life. She is the learner, the researcher, the producer, and the healer. She is intuitive and inspirational and enhanced by the business of relationships. Her chart indicates that she would do well as an industry leader in her chosen field.

After gathering all the above information Sarena began to realize how unique the combination of her gifts and talents really were. She was then able to decide how she would like to express her joy in the world (share it with others). She articulated:

"My Definite Purpose is to nurture and heal myself and others by romancing the physical senses with beautiful and fruitful products and services."

Sarena then evolved this even further into her Business Vision by adding more exact detail.

"To supply women globally with beautiful, fruitful, healing products and services, that romance and nurture their senses, and ultimately raise self-worth and self-love throughout the world."

From this starting point, Sarena created a plan for how the mechanics of her business would support this vision. She is now producing a skincare and perfume range from essential oils, using all natural ingredients, and is clearly branding these products. Until completing this exercise, she had an idea of what she wanted to do but could never decide the best way to do it. Having created a marketing plan that matches her decided Definite Purpose / Business Vision, she is able to authentically operate from her heart space.

ACTION STEP #1: CREATING YOUR OWN DEFINITE PURPOSE STATEMENT

In the book *Think and Grow Rich*[5], Napoleon Hill wrote that it only takes one good idea to become a master of success; an idea that *we decide upon* that encompasses our *definiteness of purpose*. We can uncover our *definiteness of purpose* by way of our passions, strengths, and our desired expression of these, and then we can form an *idea* around this. This will become your Definite Purpose Statement. There isn't a direct science to do this, because ultimately it is something you will describe. But there are some questions that can point you in the right direction. To get your creative and imaginative juices flowing, start by brainstorming these questions:

- What do I get really excited about doing?
- When do others comment most on how happy I am?
- What are the greatest passions in my life?
- Do I make decisions based on how much I would love to do something, or how much it will cost?
- Am I absolutely clear about my direction?
- What comes really easily to me?
- What do people come to me and ask for my help with the most?
- Am I doing what I love, surrounded by people I love? If not, what might such an experience feel like or look like?

Alisse's Example: When I started digging deeper into what I'm really passionate about I thought about things like what books and magazines I love, TV shows I watch, and precious items I spend my money on. I heard it said that if you look in your diary and checkbook you will be able to see what matters most to you. So I used this to collect information about all the things I liked and uncovered **why** I liked those things.

For example, I love watching TV shows like *The Voice* and *So You Think You Can Dance*. I'm drawn to these shows because I am deeply inspired by seeing others express their talent, tap into their inspired genius, and challenge themselves to see how far they can take their dreams.

When you list the things you are passionate about, don't leave stuff out just because it might seem frivolous to other people. What you love to watch, read, shop for, or do as a hobby are all clues!

Clarify Your Passions, Strengths, Talents, and Abilities

As you go through the questions in the section below, you can make your answers as detailed as you like. The more information you start with, the easier it is to identify patterns. You may find you complete this process quickly and easily. Or, maybe you will need to think on it for a little while longer. It doesn't really matter, just take the time you need and trust the process. When your desired purpose becomes clear it will feel right, almost like something you are remembering, that you have known deep down all along. It will be something you truly enjoy *doing*, *being*, and *sharing*. Keep persisting until you feel empowered by your result.

Client Story: I've found that the more questions you ask, the clearer your passion and purpose become. For example, one client, Doug, told me his passion was motorbike riding. However, he had no interest in competing with or selling bikes. So, I asked him if he had any other passions. He said he had a passion for fishing, which seemed totally unrelated to riding. After a few more questions he came to realize that his underlying passion was for being in nature and he loved that he rode his bike out in nature, and he loved to fish out in nature. Then, he went on to explain how he loved bushwalking and mountain climbing. He then remembered that, as a boy, he had always wanted to own his own farm. As it turns out, this client had a deep passion for nature and a love for the health of the earth. It all started to make sense and we began to form a vision around what he really wanted in his life, and how his current business was able to satisfy this.

If you keep exploring, a pattern will emerge.

Now it's Your Turn ...

What are five things you absolutely love and are passionate about doing?

1. ______________________________

____

2. ______________________________

____

3. ______________________________

____

4. ______________________________

____

5. ______________________________

____

List five of your greatest strengths and talents.

1. ______________________________

____

2. ______________________________

____

3. ______________________________

____

4. ______________________________

____

5. ______________________________

____

Now, how could you take what you're passionate about plus what you are good at and express this or share this with others?

____

____

____

____

____

____

Next, take this information and turn it into your Statement and Vision.

Start with the phrase, "My Definite Purpose is ..." and see what flows naturally after that. Ideally, express your purpose in one sentence. This is an indication that you have captured the essence of what you have decided you are here for plus it is easier to learn by rote.

Once you have this statement in place, you can be even more specific about how you desire to share your purpose with others through your **Business Vision**.

Alisse's Example: In case you're wondering, here's my Definite Purpose Statement:

"My Definite Purpose in this lifetime is to live, understand, and teach infinite purpose and potential through mastering the creative power of my mind, body, vibration, and soul."

Now you:

My Definite Purpose Statement is

____

____

____

____

____

By adding more exact details like how many people would you like to impact, the scope of this impact, and the end result you would like your clients to attain, your **Business Vision** is formed.

Alisse's Example: I have used this as the foundation of my Business Vision, which looks like this:

"To teach one million entrepreneurs worldwide to achieve true wealth, by creating a business that expresses their infinite purpose and potential, so they can experience the freedom and fulfillment they desire."

Your turn:

My Business Vision is

____

____

____

____

____

Congratulations! You have now created a **Definite Purpose** and **Business Vision Statement**. You can refer back to these for all your future ventures.

Note from Alisse: Creating my **Definite Purpose Statement** has been the most important business / life investment I have made. It has given me absolute clarity in my direction, inspired me to move forward with energy and enthusiasm, shown me the *intention for* and *way through* any of my challenges or difficulties, and has been the first step in me realizing all the abundance I experience today.

In the words of Gautama Buddha, "Your purpose in life is to find your purpose and give your whole heart and soul to it."[6] **So I believe this is the ideal place for every business owner to start.**

Chapter 2

HOW TO BECOME SUSTAINABLY LUCKY

The BASIC Principles That Boost Your Business Without Effort

As we journey through the realm of Feng Shui you'll find that the strategies you learn will make more sense when you see them as part of a complete system. In this chapter you'll become familiar with underlying principles that can help you understand the 'big picture.' We'll start at the macro level, with the Universal Laws.

THE SEVEN LAWS OF THE UNIVERSE

When you understand the three things you are here to do (as discussed in Chapter 1), your ventures will be truly worthy of the life you are spending on them. In the same way, when you understand the Seven Laws of the Universe, your ventures will be in alignment with basic cosmic principles.

The following laws provide the structure within which we have the ability to create our reality. If you aren't having success in achieving what you want, check this list to see if your actions or expectations are in conflict with these natural laws. When you work with these principles (instead of struggling against them) you will have greater flow and ease in creating and attracting the reality you desire.

The **Laws of the Universe** are very old teachings that are referenced as far back as 5,000 years ago. They can be found in many early philosophies including the Egyptian, Greek, Vedic traditions of India, as well as the Hermetic Principles of the Kabbalah. Actually, when you look closely you will find they are at the core of virtually every ancient wisdom tradition, both Eastern and Western. These Laws provide a powerful framework to guide you to work in perfect harmony with everything else.

1. **Law of Transmutation** – As we have already begun to understand, everything is energy. This energy moves in and out of visible form. Each form of energy impacts the other.

2. **Law of Relativity** – Meaning is contextual. Nothing has meaning until we give it meaning by relating it to or comparing it to something else.

3. **Law of Vibration** – Everything has an energetic vibration. Like attracts like (Law of Attraction falls in this category).

4. **Law of Polarity** – Everything has a polar opposite that is identical in nature (e.g., weaknesses and strengths). Both have value.

5. **Law of Rhythm** – All things in the Universe move in natural cycles; tides, seasons, day, night, etc.

6. **Law of Cause and Effect** – Everything you project out into the Universe comes back to you.

7. **Law of Gender –** Everything and everyone contain masculine and feminine elements. This is also called *the Law of Gestation*. Everything develops in its own time.

Each of these Universal Laws coincides with Feng Shui for your business.

Here is how:

Transmutation

This Law is a perfect explanation of the Wind and Water principle of Feng Shui. Everything that currently exists in the visible energy (material) world impacts your inner world of thoughts and emotions. Placing items, with intention, in your office space that reflect your business goals sends a positive flow of energy into your mind. Your mind then takes this energy and transmutes it back into what you create in your outer reality.

Relativity

This principle is all about perspective. When you view your current business circumstances as opportunities instead of obstacles, you are providing a positive meaning that supports you in achieving your goals. You use Feng Shui to assign meaning to objects in your surroundings to do the same thing.

Vibration / Attraction

Feng Shui assists you in identifying and changing the energetic vibration of your environment. This raises your emotional vibration and allows you to attract business prosperity, abundance, and wealth to match your new, higher level of vibration.

Polarity

Understanding and acknowledging both strengths and weaknesses makes it possible for you to decide from where you want to operate. Using Feng Shui, you can identify the functions of your business that are weak, identify the potential in them, and make them into strengths. For example, there may be a task you really dislike. This can be outsourced to someone who does like to do it, freeing your time to do what you are good at.

Rhythm

As you learn to read clues in your outer environment you will discover the ebb and flow of energy in your business. Then, you can take the right actions at the right times. For example, it may be time for the introduction of fresh ideas, or time to simply persevere with your current goals.

Cause and Effect

The energy you send out to the Universe will always return. When you pour your passion into serving others in your business, you will always be recompensed. It may not come in the form or at the time you anticipate, but it will come back to you. Feng Shui assists you in sending the highest possible vibration and energy to your clients.

Gender / Gestation

Patience is vital for business success. Set your firm intention, using Feng Shui, to create a powerful Goal Board. Then, begin taking positive actions toward your goal. They will begin germinating. Just like a seed hidden in the soil, you don't see the results right away. Let go and leave the process in the hands of the Universe. The things you are materializing will turn up in your physical reality at the perfect time for you.

FEELING LUCKY

If you've never thought much about the topic of luck, you may think it is something that only applies to people who win the lottery or always have things go right for no obvious reason. We tend to assume that some people are just lucky and there is nothing we can do to improve our own good fortune.

There is another way of thinking about luck that is much more empowering. Let's take a look at the traditional Chinese view of this phenomenon. We will explore several ways you can harness the power of your own luck using Feng Shui principles.

Three Kinds of Luck

There are three sources of fortune in your life (and in your business). Each one represents one-third of your total luck. In Chinese philosophy, these three types of luck are called:

- **Heaven's Luck**
- **Human's Luck**
- **Earth's Luck**

Heaven's Luck

This category of luck covers the life circumstances you were born into, including the mental and physical blueprint you inherited, plus the specifics about your actual birth date (time, day, month, and year). It also covers predetermined aspects of your life such as your family of origin, as well as the cultural context in which you will grow up. If you feel blessed to have been born into your country of origin, with all the advantages this offers you, thank your **Heaven's Luck!**

Human's Luck

This type of luck comes from your attitudes, habits, thoughts, and behaviors. What you decide to do with the circumstances you have been given determines what kind of Human's Luck you can expect to enjoy. This area of your luck is driven by the choices **you** make along the way.

Earth's Luck

Earth's Luck comes into play as a result of your outer environment. The positive or negative influences you encounter in your home and at your place of business can determine this kind of luck.

Tapping Into Your Luck

Understanding your luck is the first step in accessing its potential. Now that you've been introduced to the basics, here's how to start making your luck work for you instead of against you.

True Alignment

With Heaven's Luck, it is not about trying to change it but rather to align with it. You might think of this type of luck as your fate or destiny. Whether you see that as positive or negative is purely a matter of perspective.

The personality, physicality, and natural talents you were born with are your gifts, not obstacles. They are provided so that you can fulfill your Life's Purpose. Tap into Heaven's Luck by seeking out this Purpose as described in the previous chapter.

Client Story: Leni is a client who came to me after investing thousands of dollars with the best professionals and specialists for her business concept. An interior designer had helped in the office, her graphics were lovely, and she had an acceptable marketing plan and focus. However, she had no idea of her purpose in life. She had chosen her business concept from a completely logical place. Consequently, she found it to be challenging, time consuming, and lacking in satisfaction.

Once we sat down together it became very clear to her that this was not the business she wanted to invest her life into. After working through the purpose strategies described above, Leni was successfully able to sell that business and create one that aligned with what she had, deep in her soul, always wanted to do.

Self-Development

In today's 'self-help' focused age we are getting better and better at enhancing our Human's Luck. In fact, this is the most developed area of our three lucks. We understand the power of our minds and how to have a positive outlook. You have probably spent a great deal of time replacing your bad 'thought habits' with beneficial ones. You have also likely invested into good systems and the most up-to-date savvy business concepts. The results of this development may be remarkable.

However, you may find that something is missing. You still feel like you are struggling to increase your abundance in health, wealth, and success. The question is, "Are you taking advantage of the other two-thirds of your luck (Heaven's and Earth's)?" Imagine how effortless it would be to attract good fortune if you tapped into these additional resources!

Feng Shui and Luck

Here's some great news: Earth's Luck is often the easiest luck to enhance. That's because it is not only easy to see the factors that are influencing your luck, it is also easy to change them.

"If moving my desk will make me more millions then just move it."

- Donald Trump[7]

You are in control of what you surround yourself with. This includes the actual location of your office space, your furniture, the colors and décor you select. You also choose what your website, business cards, brochures, and other marketing materials look (and feel) like. The way you dress and the image you present as a businessperson are also yours to decide.

Everything in your outer environment tells a story about **you**. When you start looking at these aspects of your business, you will realize how you may have been unintentionally blocking Earth's Luck. Then, you can begin making an adjustment here, an improvement there. Gradually, you will bring your environment into perfect balance.

Watch in amazement as these small changes increase the flow of positive Chi through your business. With the help of Feng Shui, you will become a master at creating the outer environment that invites success into your life and into your work.

Client Story: Beth is a client who is a mindset guru, brilliant with making the most of her Human's Luck. However, her home office was flooded when a pipe burst and rather than fix it she simply shifted her office into the living room. After the incident she experienced clutter and crossover from personal life into business life. No matter how good her mindset, she felt overwhelmed and lacking balance. I suggested that before she did anything else, she needed to correct the office situation. She handled the issue with her insurance company and set her intention to be back in the office space within three weeks. It happened, just like that, and she quickly got back to firing on all cylinders.

Could you be stifling your business success by not implementing all three sources of your available luck? (Remember each one is worth 1/3 or 33.33% of your total business success.)

GOOD VIBRATIONS

In the Introduction we learned that 'everything is energy' (Chi) and Universal Law confirms this. Even modern day quantum physicists have now acknowledged that all matter in the universe is actually energy when you look beyond the level of subatomic particles.

All of this energy is thought to arise from the same unlimited supply that permeates the space between what we think of as particles of solid matter. Some scientists are referring to this as the "zero-point energy field." Other people use the word God, Creator, Universe, or "All That Is" when they talk about this all-encompassing Source. This is why the possibility of human potential is known to be infinite.

When you view things from this perspective, you will discover that there aren't really different *types* of energy. There are simply varying levels of vibration (or frequencies), and what distinguishes one kind of energy from another is its vibrational level.

A situation, object, thought, feeling, action, or environment that has a high vibration feels like 'good' Chi (expansive, fulfilling, loving), while one at a lower level of vibration has 'bad' Chi (detractive, destroying, fearful). We measure Chi by how something makes us feel. Of course, this is also influenced by our perception, since we decide what's good or bad as stated in the Law of Relativity. What feels expansive to you might feel detractive to someone else.

That's why it's important to listen to your own intuition when using Feng Shui to figure out how to enhance Chi in your business. Your goal is to surround yourself with things that vibrate at the level of what **you** desire to attract (or feel).

We are naturally drawn to energy that resonates with the highest level of vibration within ourselves. This is because each of us is designed to grow and expand into our highest possible level of spiritual evolution. It is our nature to achieve this potential. That's why we feel most fulfilled when we are seeking out ways to raise our level of vibration and increase our connection with the Source of all energy.

This is one reason you find yourself drawn to the works and teachings of people you admire. The attraction is about more than just the words they speak. It is the individuals who are actually living their own teachings who make the greatest impression. These leaders have a charisma (vibration) that is like a magnet which brings out the best in you. The very same reason your customers will be drawn to do business with you.

At the same time, there are some types of vibration that tend to feel consistently better than others for most people. For example, the sense of order and peace when walking into a bright, clean, good smelling office space affects pretty much everyone the same way; they feel uplifted and energized. So, you'll see a lot of Feng Shui tips in this book that work for everyone.

What Does This Have to do With my Business?

As you have discovered, your business is an outward expression of your inner state of being. However, as a complex human being, you have lots of different vibrations within you. Emotions are an example of this fluctuating energy. Some days you feel confident, inspired, and full of love for yourself and others. At other times, doubt, negative self-image, and depression may seem to take over.

Which of these attitudes do you want to express in your business branding? Obviously, you want to consistently display your highest level of vibration to your clients. That's what makes them feel good about your business. It draws them in as they feel the potential to raise the vibration in their own lives by connecting with you.

You Can Choose Your Vibration

Do you have to wait until you are feeling exactly the right way before you start thinking about your business branding and design? It never hurts to be in a creative mood. However, with Feng Shui you can take a step back from any emotional ups and downs you are currently experiencing. This approach lets you look at your branding from a higher vantage point.

From here you can see the signposts pointing the way toward a brand that sends just the right message. As everything begins to come together, you will feel creative energy and confidence in your own abilities flowing freely. Prepare to be inspired as you investigate the perfect design for your business cards, webpages, brochures, and office environment in the upcoming chapters.

Putting it All Together

Now you have the basic building blocks to understand many of the challenges and obstacles you may have been creating for yourself. Put this information to good use by:

- **Working with, instead of against, Universal Laws**
- **Leveraging all your Luck to back your business endeavors**
- **Seeking out high-level energetic vibrations**

BE A CLIENT MAGNET

Now, you are ready to start raising your business profile and profits with Feng Shui. The way to do this is by **attracting** your clients rather than pursuing them. Yes, you will network and make connections. Yes, you will still find innovative and exciting ways to introduce yourself to new clients. However, you can kiss 'hard sell' techniques goodbye.

Stop wasting energy chasing after people who don't want your services and don't feel like doing business with you. Forget the idea that you **must** have x number of leads and sales attempts to get a tiny fraction of a percentage of results. That is just more of the 'success comes from hard, unpleasant work' thinking.

Attract Your Perfect Customers

Any decent marketing professional will tell you that drawing customers to you is better than pestering them with telemarketing during dinner or spamming their email inbox. You can learn lots of good techniques for how to use the Internet and other technologies to help you make money from these experts.

Often such successful individuals have some intuitive grasp of what works and what doesn't in the way they present their business to clients. What the average marketing pro can't tell you though, is **exactly** why what they are doing works. Also, there is no guarantee that what they are doing will work for **YOU**.

That's because they aren't teaching the underlying principles, such as Feng Shui. You need to understand these fundamentals in order to build an approach that is uniquely suited to **your** enterprise, personality, and Life's Purpose. Feng Shui helps you fill this knowledge gap.

Making an Impression

Rapport is the most important determining factor in whether or not you can close a deal. What happens if customers don't feel comfortable with you (or worse, don't like you)? They don't want to engage in an exchange of energy with you; in this case, a monetary transaction.

People don't generally remember what you say. Many won't remember what you do. But, everyone remembers how you made them feel. That's why interacting with someone you just met is called 'making a first *impression*.'

We Buy Based on Feelings

We all make many purchasing decisions based on emotional impulses. We buy things that make us feel good. However, we tend to buy those things only from **people** and in **places** that make us *feel good*.

Think about the last time you considered buying something from an unfamiliar vendor or went shopping in a new store at the mall. You rely on your intuition from the start.

- How does the business make you feel overall?
- Is it orderly or disorganized?
- Are the representatives friendly or rude?

These may just be the conscious calculations that pass through your mind. You are also responding on a 'gut' level to the lighting, the colors, the layout, and everything else about the physical vibration of the business.

Does Our Outer Environment Really Affect Us?

Feng Shui isn't a 'fuzzy,' unscientific approach to making people feel good. You will be excited to learn that it is backed up by evidence. Marie Diamond, a phenomenal Feng Shui master, actually had her skills put to the test. A subject who had no familiarity with the principles of Feng Shui was connected to an EEG machine (this equipment measures electrical activity in the brain in the form of waves).

The test subject was introduced into two different environments. One room was just an average space. The other room had Diamond's techniques applied throughout. The regular room created a dominant pattern of beta (chaotic) waves in the subject's brain. This is what you would expect in a normal waking state for the average, frazzled, distracted, and overworked individual.

Exposure to Marie's Feng Shui room changed the subject's dominant brain activity pattern to alpha waves within **seconds**. Alpha waves occur when the brain achieves a flow state. They indicate extremely high potential for learning, creativity, clear thinking, and peak performance. In other words, this is the brain state people experience far too infrequently in their everyday lives.

You can read more about this test at www.learningstrategies.com/FengShui/BrainWaves.asp

High Level Vibration = Good Feeling = Instant Client Attraction

Once you grasp how positive vibrations in your outer environment can make you feel, you will realize that this energy has the same impact on your clients. That means you can create the kind of instant attraction that makes your customers feel good about doing business with you. You simply need to Feng Shui every aspect of your business that your clients are exposed to.

Imagine walking into an office to meet with a potential vendor and experiencing the following without knowing exactly why:

- You feel instantly at ease, relaxed, and confident.
- You feel more alert and aware than normal.

- You realize your ability to think and make good decisions has suddenly improved.
- You feel full of energy and ready to explore new possibilities.
- You feel that things are unfolding as they should and you are going with the flow.
- Everything seems right somehow, in balance.
- You feel excited about how this particular purchase will enhance your life.

Would you be inclined to shop with a business that made you feel that way? Absolutely!

You can do the same thing when you connect with your own clients. This isn't a trick, mind control, or manipulation. It won't work if your heart isn't in the right place. People can sense that instantly; it is part of your overall energy pattern.

So, Feng Shui is most powerful when you understand the service you are providing **and** have a real desire to tap into and meet your client's needs. Your sincere wish is that your customers experience overflowing prosperity and well-being, just as you do. You will be richly rewarded when this is your true objective.

Consider Every Point of Contact

As we've explored a little already, your customers will be attracted by marketing that is specifically designed to make them feel good on an emotional (vibrational) level. They can pick this vibration up from your business card through the colors, graphics, and other design elements.

They can sense the same things in your web design as well as the ease of navigation (flow) throughout the site. You can even Feng Shui the message on your voicemail or your hold music with the use of soothing sounds. The more you practice this art of balance, the more things you will see that can be adjusted to make them more effective.

You Are the Energy of Your Business

It is vital for you to fine-tune your business Feng Shui so that it appeals to your clientele. However, each aspect of your business must also fully support **you**. This means you will probably see significant similarities between you and your client demographic in terms of what you value in life. Like attracts like. This is yet another Universal Law that applies to your business.

This will limit the **type** (but not the number) of people you draw to your business, but only in a good way.

There are more than seven **BILLION** people in the world right now. This means 'losing' mismatched clients could only make the tiniest dent in your potential customer base. There is a virtually unlimited supply of wonderful clients just waiting for the right person to come along and meet their needs. **What a relief to access a different level of vibration, where you attract people who are thrilled that they have found you!**

ACTION STEP #2: GET CONSCIOUS ABOUT YOUR BUSINESS

Identify areas of your business where you have seen each of the Universal Laws at work (either supporting you or in opposition to you). Here they are again for your reference:

The 7 Laws of the Universe

1. **Law of Transmutation** – Everything is energy. This energy moves in and out of visible form. Each form of energy impacts the other.
2. **Law of Relativity** – Meaning is contextual. Nothing has meaning until we give it meaning by relating it to or comparing it with something else.
3. **Law of Vibration** – Everything has an energetic vibration. Like attracts like (Law of Attraction falls in this category).
4. **Law of Polarity** – Everything has a polar opposite that is identical in nature (e.g.: weaknesses and strengths). Both have value.
5. **Law of Rhythm** – All things in the Universe move in natural cycles; tides, seasons, day, night, etc.
6. **Law of Cause and Effect** – Everything you project out into the Universe comes back to you.
7. **Law of Gender** – Everything and everyone contain masculine and feminine elements. This is also called the Law of Gestation. Everything develops in its own time.

Now write your new found awareness here:

Law of Transmutation

Law of Relativity

Law of Vibration

Law of Polarity

Law of Rhythm

Law of Cause and Effect

Law of Gender

List five companies you love doing business with and why:

1.

2.

3.

4.

5.

Find one thing you can do in your own business to emulate how they make their customers feel:

Chapter 3

YOUR ENVIRONMENTAL-FINANCIAL THERMOSTAT

The Subtle Clues of Your Financial State *Hidden in Plain Sight*

In this chapter you will begin bringing your office space into alignment with your desired level of vibration. The easiest and most practical way to do this is by using Feng Shui to identify those things that don't serve you, so you can release them. Your de-cluttered office space will give you a fresh, energetically clean space to work in as you create your business branding. In this chapter you will also learn how to Feng Shui your attitude, particularly when it comes to how you think and feel about money. You may be surprised to see what's lurking around in your subconscious!

As you read this information, remember that Feng Shui works both ways. What you see, feel, taste, smell, and hear in your outer environment is not *just* a reflection of the energy you are putting out into the Universe and attracting into your life. The information your physical senses gather from your surroundings directly *influences* your mindset and can raise or lower your vibration accordingly. It's all connected.

That's why Feng Shui is such a valuable tool for helping you learn about yourself and your business. Who you are, what you do, your attitudes, beliefs, and habits are all clearly detectable in your environment. Once you learn to interpret these signs you can use Feng Shui for the next step as well, deliberately making physical changes in your environment that reinforce better habits, improved attitudes, and new beliefs about your own worthiness and abilities.

LOOK AROUND YOUR OFFICE. WHAT DO YOU SEE?

Some of the signs are pretty obvious. For example, when you have a cluttered desk or office space, this is a reflection of a chaotic, disorganized way of thinking. When you get everything straightened up, you instantly feel better.

However, many of the insights you will gain from using Feng Shui are more subtle. They only become obvious after you really start to think about why you do the things you do. Once you develop the habit of greater awareness you will notice more and more details. Here are some great examples to get you started:

Do you have a habit of never quite shutting your desk or filing cabinet drawers all the way? Perhaps you enjoy wearing jewelry that curves around but doesn't actually join into a complete circle. These are common signs exhibited by people who don't quite finish projects they start.

What about your office plants? If you have a dead or dying plant in your office, this is an outward expression of a failure to thrive. A cactus that can live for long periods without water is another problematic plant.

Water signifies the condition of your finances. You may be subconsciously expecting (and therefore attracting) a dry spell in your business.

The business cards you hand out to your customers can speak volumes about your beliefs. Are they printed on cheap material? This conveys a message, "I don't really value my business or the quality of my services."

Let's say your favorite aunt gives you a lamp that you think is really awful. Do you keep it and use it in your office because you feel like you can't get rid of it? This indicates that you have a tendency to hold on to the values and beliefs that others have passed on to you throughout your life. It's time to start doing what feels right for you.

Take a look at your figurines or other pieces of artwork around your office. Is everything a single, stand-alone piece? If you consistently stay away from decorating with items or images that are in pairs or groups, you may have an underlying feeling of alienation.

Here's another subtle clue: If there is an obstruction keeping your office door from opening all the way, you may be blocking the flow of success into your workspace.

As you can see, a lot of your habits are completely subconscious. Set internal intentions to change these. Then, use Feng Shui to imprint your intention on your outer environment by making shifts in your physical office space. See how good the resulting energy flow makes you feel!

Alisse's Example: When I first began in business I was not aware of how my environment was blocking my success. I thought that spending time assessing, de-cluttering, and decorating my work-space would only take me away from seeing more clients and generating more income. In hindsight, my environment displayed some obvious bad Feng Shui. Once I truly grasped the complexity of Universal energy I realized that this science was not 'fluff' on the top but actually the foundation of all business success. That proved to be true: I saw fewer clients and did less field work than ever before in the year I introduced Feng Shui to my business. However, my income increased to **five times** what it was the previous year. That's why I never take Feng Shui lightly.

Color Your World

Colors have a powerful effect on our moods. Each one is a different wavelength of the light spectrum. In other words, every color has its own unique vibration. The colors you like or dislike can tell you some interesting things about yourself.

Orange is a great example. Many people really don't like this color. Orange is the color that vibrates with the energy of change and transformation. If you have a negative reaction to this hue, you may also tend to strongly resist anything that takes you out of your comfort zone. It's time to become more flexible and less rigid.

Sometimes, you may feel a 'craving' for a specific color. For example, if you are in need of emotional healing you may seek out **green** spaces. If you feel like you need an extra boost of energy, wearing your favorite **red** shoes may be the perfect answer.

A little color can go a long way in many cases. **Black** has a very heavy vibration. It actually absorbs the entire spectrum of visible light. If you've ever gone through a phase of wearing black clothing all the time, you know how this can make you feel 'down in the dumps.'

We associate this color strongly with funerals and death. Use too much and it will seem to drain all the energy out of a room. The same thing can happen if you overuse black in your business graphics. However, there is a place for everything when you learn to balance it. Used properly, a portion of black can create just the effect you are looking for.

In a similar way, **white** is good in moderation. If you use this color in your business designs it indicates structure and timeliness. On the other hand, a predominantly white area resonates with a lack of care or loss of direction.

Some color meanings are culturally specific; we learn to associate a particular color with something we like or dislike. For example, if you hate **pink** and other 'girly' colors, you may be rejecting the feminine side of your nature. Remember that we all have 'masculine' and 'feminine' aspects and it can be remarkably healing to finally come back into balance. You can do this by learning to value all parts of yourself.

The Power of Images

Take a close look at the artwork in your office. This includes any inspirational posters and the image on your screensaver. Art has many different purposes. Some of the most beautiful paintings are intended to evoke feelings of longing, nostalgia, loneliness, or even sadness. Viewing these at a museum can be a powerful and moving experience; they just aren't the right fit for your office.

Here are some of the messages the art in your workspace may be sending you:

A train that is headed straight for you. That's a pretty aggressive image! It can make you feel like you are under threat. In fact, this oncoming train is a good example of the type of symbolism that Feng Shui practitioners call a 'poison arrow.' We'll learn more about how to avoid this specific type of negative energy later on.

What about the opposite kind of image? A desert island with a single palm tree would be a nice, peaceful scene. There's just one problem. It's deserted. There's nothing in this image to suggest connection with others, only isolation. Unless you have some previously undiscovered business model that doesn't require clients, other people are still an essential component of your business! Your décor needs to reflect this.

Some images can be positive or negative depending on their location in your office. For example, a mountain that is hung on the wall in front of your desk is a giant obstacle. Placed on the wall behind you, it becomes a source of support and strength. It can even be a reminder of how far you have come.

A beautiful seascape with a ship tossed on rough seas indicates turmoil. An empty sailboat signifies loneliness. Any image with an object or animal moving away from you indicates a quality that is leaving you. For example, a ship sailing into the distance would indicate that your wealth is slipping from your grasp.

Trees can be a powerful image of growth and life. But a picture of a tree that has dropped all its leaves is dormant, not bursting with vitality. The green of spring and the abundant foliage of a tree during its time of greatest growth would be a much better choice.

Inspirational posters are a favorite selection for office décor. Unfortunately, some of the images used send a message that is in conflict with the purpose of the poster. For example, a rock climber dangling in midair might be seen as an intrepid adventurer. However, you don't want to feel like your business venture is hanging by a thread!

Habits and Symbols of Lack

Are you the kind of person who feels naked if you aren't wearing your watch? Take a look around your office. You probably have clocks and calendars everywhere. You are very conscious of time and feel driven by deadlines. Maybe you feel like you are running out of time or believe you don't have enough time to finish all your projects.

Sometimes, people demonstrate a belief in lack by choosing certain types of decoration for their office. A collection of empty pottery is a good example. If you find yourself consistently drawn to 'empty' items, perhaps there is something missing in your life.

Do you hang on to stuff like old business brochures because you can't bear to part with something you paid for? Some people hoard whole boxes of business cards with outdated phone numbers on them. Being a packrat is one very common outward expression of a belief that there 'isn't enough.'

If you are resistant to letting go of things like this, you have a subconscious belief that there isn't more money coming in. As long as you keep hanging on to the old, you won't be able to progress forward. Recycle those old brochures and move on!

Blurred Boundaries

If you work at home, the most obvious boundary you can set is the one between your office space and the rest of your house. You can do the same thing if you work outside the home by practicing little habits that demonstrate your intention.

Shut the computer down at the end of the day. Also, clear all the clutter and any files from the surface of your desk. This will give you a fresh start each morning.

Place a picture of your family on your desk in the top right corner. This will remind you to find a good balance between work and family time.

Where's Your Focus?

Keeping your visual field free of negative energy is critical in Feng Shui. Whatever you look at all day long is what you will bring into your reality. Quick, go check your office. Are there bills left out where you can see them? This is a constant reinforcement of money flowing **out** of your business, not into it.

Alisse's Example: I used to keep all my paperwork in open slots; I thought this would make it easy to access everything. However, I always felt pressured and overwhelmed when sitting at my desk. It was like an open 'to-do' list sitting in front of me. My bills, receipts, and bank statements were also in there with all the other paperwork. So, of course, I was always overcome with debt. The change was instant when I had doors added to the shelving and was able to remove these things from my line of vision.

Put all your bills away in a green folder, this represents money and abundance. Try adding a slip of red paper (that stands for confidence and action). Write an affirmation on the paper. "I pay my bills on time every time," is a great positive statement.

If you need reminders to pay your bills, set up automatic prompts in your computer calendar. Then, don't think about bills unless you are actually in the act of making a payment. There's no need to think about them the rest of the time. This just draws your energy away from creating the prosperity that will allow you to pay your bills effortlessly.

Proper Placement

Do you feel like you are butting your head against a wall all day? This may be because you have actually placed your desk directly up against a wall. The feeling is compounded if you don't have a window in that wall.

Are you tense and on edge? The placement of your desk can make you feel this way if your back is to the door. Your natural response is to feel nervous or like you may be ambushed at any moment. We will discuss the ideal way to arrange your office furniture later on in the book.

Are Negative Anchors Dragging You Down?

Feng Shui is all about raising the vibrations of your inner and outer environment. For humans, vibrations are expressed in the form of emotion. We attach meaning to objects in our environment all the time. Nothing has a meaning until we assign one. We label objects and anchor emotions into them. For example, you might feel a creeping sense of dread whenever you turn on your computer and prepare to open your email inbox. There is nothing inherently positive or negative about your computer or email. They just are. This means you can choose to feel about them however you want.

The first Feng Shui step you might take to change how you feel about turning on your computer in the morning is to make the screensaver a beautiful picture that inspires you. Sunflowers represent golden opportunities and make a great 'wake-up call' to start your day. Or, you might choose to have your Goal Board with all the wonderful experiences you are attracting into your life as your screensaver. These are tips for improving the vibration of your outer environment so what you see instantly uplifts you.

On the *inside*, you need to figure out why you have anchored a negative emotion into a particular item or activity. Often, this is based on a limiting belief you hold in your subconscious mind. For example, your dislike of accessing your email in the morning might stem from a belief that people only contact you to complain. Once you have identified this belief, you can create a positive affirmation to anchor into your email. So, each time you click your mouse to open a new email, you might say, "I value feedback because I love how I feel when I do a great job for my clients." This can change the anchor you have attached to opening your email from one of dread to one of empowerment. You can also use visualization to imagine opening a dozen emails that are full of praise from satisfied clients. Follow this up by making a folder and storing all the positive feedback you receive, then open this up and read it first. Now doesn't that feel better?

Alisse's Example: I had a very expensive, handmade mirror given to me. It was a reward for reaching a level of success in a previous career. Unfortunately, when I decided to leave that business, it was not with good memories. I had the mirror hanging in the entryway of my new business. The challenge was that the mirror did not represent my new business. It was a symbol of an old business and a reflection of negative emotions and feelings of failure. I sold the mirror (even though I loved the look and style of it). The minute I took it down I found another mirror to hang there that I loved just as much. By the next week, I had been offered a new opportunity, which has led me to all my current success.

CREATE NEW ANCHORS AND NEW HABITS

You can do this inner and outer Feng Shui for anything in your business that makes you feel resistant to moving forward toward success. For example, you can program your ring tone on your mobile phone to play a cash register sound, "Ka-ching!" to represent money coming in. Then, you can even use a positive affirmation in the way you answer the phone, "It's a great morning at Innovative Business Solutions. This is Kathy. Tell me, how I can make your day today?"

What if you drive a lot in your business? Perhaps you really don't like the way it feels to get in the car because you believe you will be exhausted by the end of the day. Buy yourself a beautiful keychain so you see it every time you unlock your car. Put in a CD with music that is sure to make you smile as soon as you turn the key in the ignition. Try playing

I'm Walking on Sunshine[8] to start your day. Or, pop in the soundtrack from your favorite dance movie.

Every good or bad habit you have is preceded by an anchor of some kind. For example, you might have a file tray on your desk that acts as an anchor. When you see it, you are triggered to stack your paperwork in the tray. However, you don't yet have a habit of sorting the papers into the filing cabinet. So, the paperwork just stacks up in the tray until it is overflowing. The tray has become an anchor which creates clutter and disorder in your office space. It is time to take everything out of the tray and file those papers away where they belong. Now, place the file tray in a desk drawer or cupboard where you cannot continually see it. Make a sign for it that says: "I File All Paperwork by the End of Each Day." This creates a new anchor that triggers you to develop the habit of keeping your paperwork in order and reducing clutter. Habits are always easy to maintain once they are ingrained. So, take advantage of this fact by building habits you want to keep.

Clearing out old anchors to make way for new ones can be as simple as cleaning off your desk completely at the end of each day. That way you start with a blank slate in the morning and don't fall into unintentional habits or start attaching emotions to unfinished work at the end of a challenging day. The Feng Shui vibration of a clean desk says, "I have a fresh start and unlimited possibilities!"

MONEY, MONEY, MONEY

Before we complete this chapter with fun and energizing action steps to de-clutter your office, we need to take a quick trip through your subconscious. It's time to clear out the cobwebs of old beliefs about money that may be holding you back from success. Then, you can invite in even greater prosperity by making positive changes in your environment. In a later chapter you'll even get some exciting tips on how to Feng Shui your finances!

Accessing the Wealth You Deserve

How you value yourself is often directly reflected in your level of financial success. Why wouldn't it be? Money is a form of energy that you can attract or repel just like any other. When you raise your sense of self-worth, love yourself, and understand that you have immeasurable value, you attract wealth more and more easily.

This doesn't mean you think money is more important than anything else. On the contrary, you see it as simply one of many types of wealth, abundance, and prosperity that you deserve to enjoy in your life. This includes vibrant health, satisfying personal relationships, and much more.

Seeking True Prosperity

When you understand the nature of true prosperity, your approach to everything in life will be more balanced. You realize that money serves you, not the other way around. It is a means to an end, not an end in itself.

After all, if you were stuck on a desert island with a suitcase full of cash it would serve no purpose. Money only has value within the framework of society. It is part of each culture's mutual agreement about money's energetic significance.

Abundant financial success simply facilitates your ability to enjoy the things that are actually important in life, such as:

- Having a career that allows you to spend your days doing what fulfills you most
- Being able to devote lots of time to simply being with your loved ones

- Getting to explore and experience the world and all it has to offer
- Experiencing health and wellbeing that make life more enjoyable

GETTING OVER YOUR FEAR OF WEALTH

We tend to have a very complex emotional relationship with money. Many of the facets of this relationship are fear-based. For example, we may be consciously or subconsciously afraid that:

- **We don't have enough money now.**
- **We won't have enough in the future.**
- **We don't deserve to have or make more money.**
- **We don't have the skills / abilities to make a lot of money.**
- **We will lose the money we have accumulated at some point.**
- **The only way to get enough money is to 'spend' all our time earning it.**
- **Wanting more money means we are greedy and materialistic.**
- **If we had lots of money we would become selfish and horrid.**
- **If we start making more money than our spouse he / she will feel like a failure.**
- **If we start making more money than our friends / family they will resent us.**

That's just 10 of the most common fears we have about money. These beliefs limit how much money we attract into our lives. You probably personally identify with at least three to four of those fears immediately. If you dig deeper, your personal habits may reveal that you secretly hold even more of these negative beliefs than you originally thought.

Alisse's Example: One of my biggest challenges was poverty consciousness. I grew up without a lot of money and developed many fears around it. I worried a lot about not having enough. I did not realize how much like this I was until I noticed how many things I held on to because they cost me money to get in the first place. They had become useless to me, but I had a desire to keep them anyhow. This was especially true for clothing from business 10 years ago. The garments were good quality and cost a lot at the time, though I was never going to wear them again. There were many other signs showing my beliefs and fears about money as well. I hung on to old business cards with outdated information. I had my bills out in the open where I could see them. I had a picture on my screensaver of an empty boat floating away into the distance. My wallet was a mess and it was old and black in color. Using Feng Shui helped me uncover and acknowledge these hidden beliefs so I could take control of changing them.

Examining Your Self Worth

How about the idea that you don't deserve to make more money? You may think, "Of course I deserve to make more than I do now." But what do you feel inside at the thought of raising the price of your services by 10%? How about 20%? How about 100%? Forget for a moment about whether or not you *believe* your client base would pay those prices.

At what point would you start feeling uncomfortable with how much you are charging? That's the cutoff point where you are telling the Universe, "Don't send me clients who value my services above x number of dollars per hour."

This isn't saying you need to go out tomorrow and double your pricing. Raising your prices gradually is generally a better approach. It gives you time to start attracting a new client base to replace those who don't value your services highly. However, you might start rethinking *right now* how much you really believe your services are worth on an energetic level.

Remember, money is a form of energy and payment is a form of energy exchange. If you are pouring your precious time, your boundless enthusiasm, and your sincere passion into what you offer your clients, your services are worth a great deal. This applies whether you are a dog trainer, a cosmetics saleswoman, a legal advisor, or a tennis coach.

As long as what you are offering comes from your heart and is in alignment with your Life's Purpose, it is filling an important need for the clients you attract to your business; otherwise, you wouldn't feel drawn to doing it.

Alisse's Example: When I hung my certificates on the wall in my office I immediately felt more confident in the service I offered my clients. I always had the certificates. In fact, they were stored right by me in my desk drawer. Though the minute I put them up on the wall in lovely matching frames, I felt more worthy and more valuable (and my clients started to say the same).

Is it Greedy to Love and Appreciate Money?

You may have been taught that it is greedy to seek prosperity and abundance. Perhaps you believe that self-sacrifice is the only way to ensure others have enough. This is backwards thinking! You can only **give** from what you **have**.

The creation of abundance results from a continuous cycle of giving to yourself *and* giving to others. Provide abundantly for yourself; then you will have plenty to share with those around you. When you give generously, you are rewarded in kind. The energy you send out comes back to you (this is one of the Universal Laws you learned earlier).

Do you still think you shouldn't enjoy prosperity while others around you are struggling? Remember, you have the greatest degree of direct influence over your *own* life. Use it to show those around you what is truly possible. They have the same power within themselves, if they choose to access it. Setting an example of what can be achieved is the best way to positively impact the lives of others. Don't hold back; lead the way!

The Prevailing Cultural Myth About Success

Most of us were raised in cultures that place a great deal of significance on **hard work** as the key to success. When you take a look at the mindset behind that terminology, you will notice something interesting.

It assumes that 'real' work, the kind that will help you achieve your goals, is defined by struggle and difficulty. If you aren't working as hard as you can, this is attributed to a lack of self-discipline (*note that discipline is a word we often associate with punishment*).

Since your attitudes and beliefs play a huge role in the type of situations you attract, you can see how this would become a self-fulfilling prophecy. On a subconscious level, you have been taught that success ***requires*** struggle and difficulty. All too often you believe this without question.

So, that's exactly the type of energy you attract into your business ventures. You work hard (often every day with little or no time off). Then, when you get worn out and feel like giving up, you think there is something wrong with your strength of character! You believe that if you only had more self-discipline you would have more success.

Releasing the myth of hard work is a critical step on the path to real prosperity!

Alisse's Example: I have always had the underlying programming that life and success is hard. It is not what I wanted to believe, it was just the way I had learned to think. The picture I had hanging in front of my desk was of a house sitting on top of a large hill with a long path climbing up to it. I loved this picture. It was a nice print. But it represented a long climb up to the house indicating that I expected things to be hard in my business. Once I replaced this print with my Goal Board I started to notice how much more I enjoyed sitting at my desk. And to my great surprise those things I pasted on the board began to simply turn up in my life.

When you let go of the myth of hard work, you will begin to have different expectations about how work **should** make you **feel**. You will expect it to be enjoyable, inspiring, and fun. Remember, your beliefs, intentions, expectations, and emotions (vibration) set the stage for what will actually create in your life.

If you start feeling like work is a struggle, look and see what is blocking that easy flow of energy (Chi). This might be something in your internal belief system, something in your external environment, or both. I'm guessing you've heard the phrase, "Work smarter – not harder." Tapping into the energy you need is much smarter than just trying to 'muscle' through difficulties with sheer will power.

Don't feel bad if you've bought into the cultural myths about money. You're very capable of changing these beliefs and inviting prosperity! Let's start by clearing out all the anchors in your office that may have reinforced old attitudes and habits.

ACTION STEP #3: CLEAR YOUR SUCCESS PATH

Set aside time today to sit in your work environment and take a good look around. Ask these questions about the overall space:

- Do I feel good sitting in this space?
- Does my energy feel replenished, or depleted?
- What does this space say about me and my business?
- Does this space support or hinder my business success?

Then look at each object individually and ask these questions:

- What is the origin of this object?
- What story does it tell me? (Remember that all objects hold memories, feelings, and associations.)
- Does it support or hinder my business success?

De-clutter Your Environment

This is a vital initial step in the Feng Shui process. Trying to improve the Chi of an environment that is full of clutter is about as effective as drinking a glass of water with a hamburger and fries to improve your health.

Note from Alisse: I found it very difficult to de-clutter when first introduced to Feng Shui. In fact, I would have to say it was the most challenging thing I faced. I could always think of a reason why I needed to keep something. You know, the 'one day I'll really need this' scenario. However, the truth was that this stuff just sat in my space and caused clutter and confusion. As a result, I could never decide exactly what direction I was going to take my coaching and I often overwhelmed my clients with the same lack of clarity I experienced.

Getting my office in order has made a huge difference in my ability to stay focused. Even today, I still de-clutter regularly!

1. The most effective way to de-clutter is to move everything out of your entire office space, especially from storage areas. Then, only bring back in what is relevant (those things that you need, use, and love).
2. While doing your entire office in one session is a very cleansing approach, it may be too overwhelming. So divide it into sections and make a de-cluttering schedule. Start at the doorway and move around the room in a clockwise direction. You may be able to manage one section per day or one per week. The sooner you can get it all done, the sooner you are clearing the way for new energy and new opportunities.
3. Sort items that you do not want into piles for the rubbish, for repair, for recycling (sell or donate), or for return (if they belong to someone else). Mark it in your schedule to remove these items from your premises as soon as possible. Don't just relocate them to the garage or storage shed unless absolutely necessary. If you do store unwanted items, make this a short-term solution and be sure to move them as soon as you can.
4. De-cluttering can be challenging. Remember that your outside environment is a reflection of how you feel inside. We hold on to things for many reasons. Be patient with yourself and break this process down into bite-sized pieces. But don't avoid de-cluttering just because it is challenging. If it is just too overwhelming, hire a professional 'de-clutterer' to assist you. There is nothing wrong with getting some help.
5. Once you complete the initial de-clutter, maintain it by doing 15 minutes of cleaning and organizing each day. You will soon

realize that it actually feels quite nice to do this 'cleansing procedure.' It's a great way to end each day since you will come into a clean, fresh space in the morning to begin work.

PART TWO

Chapter 4

THE POWER OF INTUITION IN BRANDING

Calculating Intensity – Combination – and Elemental Energy to Create Your Profitable "High-Energy Message"

Now that you've cleared out your office, it's a perfect time to get clear about your business branding. This includes your logo (but that's just the start). Your brand is one of the most important ways you communicate the energy of your business to your current and future clients. Think about it this way: If you hire someone to go out and promote your business, it's essential for that person to be sending the right message about you and the value of your services. In the same way, you want to use branding that sends a targeted, accurate, high-energy message to attract your perfect customers.

In the last couple of chapters, you've begun to think about the true nature and value of what you are offering customers. Before you move forward with branding, we want to get very specific about this. That way you are branding based on the *core* of your business and not a passing fad or a surface aspect that isn't essential.

Another critical point of branding is knowing *whom* you want to attract as a customer. You must know what they desire or what they see as a problem, so you can accurately describe your solution.

Alisse's Example: Here's how I articulate my **Brand Message**:

My service is: Business Coaching and Designing

The value is: I am assisting clients to increase their wealth and success by building conscious business models, deliberate energetic brands, and harmonious workspaces.

The unique value: I use 'energy-based modalities' and 'Universal principles' combined with 'savvy Western business strategies' as the basis of my Coaching Systems.

(This is your **basic unique selling point,** or what differentiates you from other businesses of your kind.)

The unique purpose: I am deeply passionate about assisting clients to design a business and a life that expresses their definite purpose and infinite potential in a highly profitable manner.

(This unique purpose comes from your 'Definite Purpose.' Because it is based on **your** deepest passion, **your** unique strengths and talents, and the way **you** desire to individually express this to the world, this is your **most unique selling point.** It is not something that anyone else can duplicate and it is the key to your business success and sustainability.)

My target customers are: Entrepreneurs (aged between 35 – 55 years).

Their challenge: These individuals are people who are tired of having no time, no balance, strained relationships, strained health, no meaning, no purpose, who are trading their life for money alone and no longer want to do this.

My solution (this comes directly from my Business Vision): I teach my clients how to master their personal and business vibration so they can create a business that expresses their infinite purpose and potential, and experience the freedom and fulfillment they desire.

Business Vision Restated: To connect one million entrepreneurs worldwide with true wealth by having them create a business that expresses their infinite purpose and potential, so they can experience the freedom and fulfillment they desire.

This information is your core Brand Message. It is the basic information you will use to choose the words, shapes, colors, and designs that most accurately align with your business. At the end of this chapter you will have the opportunity to create your own Brand Message (or check off your existing one using the Feng Shui fundamentals).

OK, let's get started by looking at specific examples of Feng Shui for your business branding!

FENG SHUI YOUR BRANDING

In today's global business environment you may work with a lot of people you never see face-to-face. This means your branding reaches them mostly through your marketing material, website, and email correspondence.

How do you build trust with people based on what they see of your business within a few seconds of entering your website? After all, you can't offer to make them a cup of tea and sit down for a nice chat so they can feel comfortable with you. They won't see how beautifully your office is decorated or that you are wearing your power suit (although these things are still very important).

The most critical characteristic of your brand can be expressed in one word: **consistency**. All your marketing materials, emails, and other correspondence must be consistent in the following areas:

- The colors you use and all other design features (logo, style, etc.).
- Your unique selling point (what you offer your clients that no one else can).
- Your business name and tagline (this expresses your vision and purpose for your business).

Consistency gives clients an instant feeling of trust and confidence. When you present a consistent image you come across as a professional they can depend on to follow through and deliver on your promises. It is obvious to the customer that you have a clear direction in your business and that you believe in yourself.

When you set a clear intention to be consistent in your business branding, you also build your own confidence and self-trust. That's another excellent reason to really dig into the message you want to convey in your branding. In a very real sense, it defines your identity.

Accurate Branding

That brings us to the second vital thing about your business branding, **accuracy**. Obviously, you don't want to give your customers the wrong impression about what you do. You also don't want to miss out on the opportunity to make the right first impression.

Ideally, the name of your business tells people what you do.

For example, if your name is Amy and you are an interior decorator, you might be tempted to use a name like **Amy's Visions**. However, this doesn't tell your potential customers what you have visions ***of***. A business name like **Amy's Interior Designs** would come closer to the mark. Then, you can use the term vision in your tagline, e.g.: "**Making your home a vision of comfort and style.**"

This combination of business name and tagline is powerful because it tells people:

- **Who you are**
- **What you do**
- **What you offer them**

See how easy that is? Often, creative people like to get clever with their business names. This doesn't hurt; however, it won't do you any good if no one 'gets' the reference or pun you are making. You can still include enticing **wording** in your business name and tagline (even quirky or quizzical if you like) as long as you keep your overall message clear and accurate. Good communication skills are something your clients will instantly appreciate and understand.

Tips for Creating a Great Logo

1. Start by making a list of possible ideas for your logo. Some options include your whole business name (like Google), the initials of your business name (like HP), a shape or lines (like Adidas), or a symbol (like Apple). Don't limit yourself in any way at this point. Just start jotting down concepts as they come to you.

2. Make sure your logo is an accurate and relevant representation of your business. For example, Mercedes wheel spokes elegantly create a very industry and product-specific logo.

3. Keep playing with concepts until you come up with something powerful / impacting / interesting / dynamic. Too many people pick generic and boring logos. They miss the opportunity to represent themselves well. There are so many creative things that can come from shapes, lines, colors, flowers, animals, games; our world is full of symbols.

4. Move toward simplicity. A good artist can streamline your initial logo symbol / idea into something very elegant, refined, and meaningful. Think how simple the Nike 'swoosh' is! That checkmark is the symbol for getting things right and for getting things done. The goal of your logo should be to make your client feel a specific way. Later on, we'll look at a wide variety of shapes and symbols and what feelings they generate in the viewer.

5. Don't forget that a logo can be a very powerful tool to tap into your customers' subconscious. Yoga Australia is a great example of how to do this with subtlety and style. They use the image of a woman holding a yoga pose. When you look again, you can see the country of Australia in the gap between her body and the position she is holding her leg. That's a perfect example of using a logo to subconsciously make a connection with your customers.

6. Get ideas for your logo by exploring other business logos. Find five you love and five you dislike. Articulate what you do and don't love about these logos. This is a good list to take with you when visiting a graphic artist, since they work well with visuals supported by detailed explanations.

7. Make sure you know the exact meaning you intend behind your logo. Place your intention on the logo deliberately with complete understanding. This energy will be projected out to your clients and thus attracted back to you.

Client Story: Chris is a client who came to me for help with his logo. His business is Strategic Business Planning. When he first presented me with the logo drafts his graphic designer had prepared I was shocked at how generic they were, not to mention how much he had invested in them. Chris prides himself on offering a service with a real difference, something innovative and edgy. Yet his designs said 'conservative' and 'mainstream.' After talking, Chris decided that he liked the idea of a chessboard. He chose the Knight as his symbol because he felt the piece best represented his service, and he personally loved horses. His finished logo looked incredible. It was edgy and innovative, just like the services Chris offered.

Would an innovative and interesting logo better represent your brand message?

When I Feng Shui a Logo I Look For These Things

- Does the logo catch my eye? It needs to say, "Look at me / Pick me up / Keep me!"
- Is it relevant, does it mean something? There is an agreed meaning attached to most symbols. For example, cats are independent and represent confidence. Keys open something that is locked.
- Does the logo look strong and purposeful? It needs to, or clients will feel you are not reliable and professional.

- Does the logo contain any pointed or sharp edges? You don't want anything pointing at your business name; this will harm your business success. The same rule applies for anything offensive or perceived as dangerous. These symbols will repel the customer and make them wary. For example, scissors used by a hairdresser or a knife by a chef are poor choices.
- Does the logo have any missing parts? This will make clients feel like you are not complete in what you offer. Wikipedia's logo is missing some puzzle pieces; think about what that says to the viewer regarding what's lacking in their site!
- Is it unique and creative but still very clear? Clear designs with no distractions represent that you know what you are doing. Uniqueness means your business is innovative (not generic) so you stand out from the crowd.
- Are the colors sending the right message and are they relevant? Are they the right intensity, the right combination, complementary of your message and each other? We'll talk about colors in more detail later.

Alisse's Example: This symbol was very specifically designed for one of my business arms *SUCCESS CREATION COACHING AND SEMINARS, Tag Line: "Create Your Inner Connection With Self, Health and Wealth."* The dollar sign represents prosperity and it sits over a circle. Notice it is not trapped *in* the circle, which would limit growth. The symbol of the dollar sign has turned the circle into a yin / yang symbol which represents the connection between the 'inner self' and the 'outer self' and achieving balance.

The outer edge is gold and represents a gold coin (treasure). I have enormous respect for the message shared in *The Secret* movie[9] and the industry professionals that were featured in it. So, I wanted the gold to be the very old orange gold in color with the modern feel of the new, brighter, more innovative colors in the centers. *The Secret* used the symbol of the S in a stamp and I wanted to have mine as a stamp or seal as well.

The S in my dollar sign works nicely for the word Success also. Naturally, the gold represents wealth. The deep purple on one side of the 'S' represents power, wisdom, learning, and belief in self. The green on the other side is about growth, the color of money, and the heart. This symbol has a lot of energy for me. It is a complete message that I am able to project out into the world. It makes me feel good when I share it!

See how comprehensive you can be with your logo and the message you send to your market.

INTUITIVE BRANDING WITH COLORS AND SHAPES

The colors and shapes you use in your business brand send a message about your business. It is time to look beyond the surface meaning, to see how this part of your business branding makes your customers feel. Also, when you have a graphics professional do your designs without much input from you, the result is unlikely to reflect your true brand.

How can you correct this? It is time to listen to your own intuition. Forget about what other people in your industry are choosing in terms of design and color. We now know that your business is unique.

Use Feng Shui as a framework or structure within which to explore your intuition. The principles of Feng Shui reveal the underlying meaning and feeling that resonates within each shape and hue. This makes it a great source of inspiration. As you learn more about colors, you will find the perfect combination that expresses the purpose and vision of your business.

Reach Within to Find Your Perfect Colors

As you start choosing colors, trust your own abilities. Use a positive affirmation such as "I intuitively know the perfect colors for my business and will recognize them when I see them." This will help you feel confident and set your intention to make just the right choice.

Take your time and take a break if you start to feel overwhelmed with information. Often, if you can't make a firm decision it is a good idea to sleep on it. The answer may come to you in your dreams or appear obvious in the fresh light of the morning.

If you find you are still completely unsure which colors are best, don't be afraid to seek help. A consultation with a professional trained in Feng Shui Design can help you clarify what you are trying to say and what feels right for your business. Plus, it never hurts to get a second opinion.

Color Chart:

- **Red:** confidence, leadership, strength, drive, will, power, speed, life force, action, energy, physical passion.
- **Pink:** compassion, endearment, support, acceptance, femininity, unconditional love, self-love, service, community care.
- **Fuchsia:** Universal love, passionate love, fun, self-expression.
- **Purple:** luxury, wealth, royalty, prestige, honor, hope, self-belief, spiritual learning and understanding, inner strength, faithfulness.

- **Orange:** stability, assurance, nourishment, appetite, nurture, creativity, joy, happiness, worthiness, pleasure, playfulness, flexibility, freedom.
- **Yellow:** clarity of thought, intelligence, learning, mind expansion, concentration, decision-making, memory, clarity of direction, courage, positivity.
- **White:** purity, simplicity, silence, divinity, completeness, precision, structure, cleanliness, timeliness (overuse feels isolated, cold, and lost).
- **Gold:** prosperity, ultimate clarity, wisdom, enlightenment, higher thinking, optimism, radiating glow, true wealth, purpose.
- **Silver:** completion, intuition, magnificence.
- **Blue:** communication, flow, dependability, sincerity, friendship, relaxation, serenity, calm, devotion, trustworthiness, authenticity, empowerment.
- **Turquoise:** being true to yourself.
- **Dark Blue:** responsibility, ambition, authority.
- **Black:** mystery, elegance, sophistication, strength, protection, secrets (overuse is depleting).
- **Green:** growth, abundance, new ideas, freshness, health, healing, balance, forward movement, harmony, emotional strength, the color of money.
- **Olive Green:** patience in growth.
- **Brown:** grounding, earthy, sensual, rich (overuse feels stuck, bogged down).

Which of these colors resonate with you? Which align with the solution you intend to offer your clients? Which best suit the overall message of your Brand?

Think about how you want to make your clients feel. Perhaps you are a financial advisor offering a service that promotes your client's growth and abundance. The color green reflects those qualities and might resonate with you and with your customers.

Suppose your expertise is in long-term financial planning for younger people who are decades away from retirement. Would olive green (patience and growth) be more suitable? Maybe blue would be ideal. After all, you are asking your clients to trust you with their financial future.

As you can see, interpreting the right colors for your brand requires a deep understanding of your own business. That's a good thing. Clarity only enhances your success.

Also, this phase in the design process can be a lot of fun! Play with the different colors and combinations of colors without stressing out about making an immediate choice. The answers are there and will come to you at the right time.

Color Combinations

Using two colors in combination allows you to convey a more complex message than just one color by itself. You may find yourself drawn to more than one color on the chart for a reason. Let's explore a few of the different color combinations and what they mean to your customers.

Browse these combinations to get an idea of how they work in creating a message. Then, figure out how your favorite color combo will make your customers feel.

- **Red and Green:** an action orientated business that sends a message of growth and getting the job done.

- **Purple and Pink:** an inspirational business that can be counted on for the 'Care' factor.
- **Orange and Purple:** a business that nurtures your self-belief and worthiness.
- **Red and Purple:** a focus of success and connecting business spirit with physical action.
- **Pink and Yellow:** a business that says, 'I know what I do and I intend to share it with love.'
- **Red and Yellow:** radiating positive action-taking.
- **Purple and Green:** a business interested in setting and achieving goals.
- **Orange and Pink:** celebration of life.
- **Pink and Green:** a business promoting the message of love and romance.
- **Yellow and Purple:** knowing and having faith in who you are; a business promoting the enhancement of self-esteem.
- **Gold and Turquoise:** a business interested in the flow of true wealth.
- **Turquoise and Green:** a business wanting to enhance feelings of relaxation and ease.
- **Blue and White:** a professional business with a clear plan to get results.
- **Blue and Green:** a business speaking trustworthiness and sustainability.
- **Blue and Gold:** a business promoting financial success.

Lucky Number 3

While two colors can look sophisticated, it is considered especially auspicious (lucky) to use three colors in your business branding. This doesn't mean your design has to look loud and overpowering (unless this suits your brand message). It may be appropriate to add just a touch of your third color, as an accent for your branding.

It is also considered auspicious for a brand to contain all five of the elemental colors (as you will learn about below); however, this needs to be carefully considered and cleverly designed otherwise it may have the complete opposite effect. Have a professional help you with this if you think it is suitable for your business type.

ELEMENTS AND BRANDING

In Eastern philosophy (as in many cultures) the world is seen as being made up of several basic elements. These aren't the ones you learned from the periodic table; they are concepts drawn from the visible natural world. These elements are Earth, Metal, Water, Wood, and Fire.

Each element is associated with its natural material, as well as a set of specific qualities as we experience them in our environment. In addition, they are linked to specific colors and shapes. Here is a basic example of what we're talking about:

	**EARTH**	Yellows and Oranges	Gathering and Nurturing
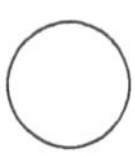	**METAL**	White and Metallic	Solidifying and Structuring

WATER	Blues and Black	Flowing and Allowing
WOOD	Greens and Browns	Generating and Sustaining
FIRE	Reds, Purples, and Pinks	Activating and Expanding

Different types of businesses also relate predominantly to one of the five elements and are therefore enhanced by certain colors and shapes. Sometimes the link is obvious. For example, a pottery artist who uses clay is an Earth-based business and would be complemented by the color yellow or a logo based on the symbol of a square. A business that makes timber furniture is Wood-based and can benefit from the color green or a logo based on the rectangle shape.

However, for other businesses we need to look further into each element's qualities to see which one relates best.

ELEMENTAL QUALITIES FOR BUSINESS TYPES

Earth Business: grounding, centering, supportive, securing, informing, nurturing, harmonizing, analyzing, nourishing, resourceful.

Metal Business: systemizing, focused, clear, planning, structuring, organizing, orderly, refining, completing, hygienic.

Water Business: empowering, thoughtful, flowing, serene, enlightening, trustworthy, devoted, faithful, authentic, loyal.

Wood Business: growing, developing, healthy, healing, fresh, forward-moving, uplifting, raising, sustaining, achieving, goal-oriented.

Fire Business: action-oriented, expressive, expansive, energetic, confident, vibrant, enthusiastic, illuminating, inspiring, passionate.

To determine the elemental nature of your business, choose the list of qualities that best describes your business considering the product or service you offer. Again, refer to your Brand Message to help you identify this.

Once you have decided the type that best represents your business, check if the colors and shapes in your logo complement it. For example, a fitness trainer could work with the fire element (reds and triangular shapes). A nutritionist might choose the wood element (greens and rectangular shapes). A dentist could select the metal element (white / metallic and round shapes).

Here are some more examples of business types that relate to each of the five elements:

Earth-Based Businesses:

Property, Earthmoving, Research, Interior Design, Beauty, Science, Childcare

Metal-Based Businesses:

Law, Business, Metal Sculpting, Technology, Engineering, Architecture, Dentistry

Water-Based Businesses:

Swimming Pools, Fishing, Shipping, Travel, Glass Making, Sales, Beverages

Wood-Based Businesses:

Fashion, Furniture, Flowers, Stationery, Construction, Nutrition, Healing

Fire-Based Businesses:

Finance, Sports, Electricity, Entertainment, Bright Objects, Cooking, Intimacy

Picking the right element is mostly a guide to enhance your business. It is not intended to limit your message or how you create your brand. However, the elemental system does become crucial to understand when it starts working *against* the flow of your business. Here is an example of how this can happen:

Watch for Elements in Conflict (Hidden Destroyers)

Pay special attention to whether the colors and symbols you have picked might unintentionally weaken or control your business. You would be best to limit or avoid a color or shape that is *not* complementary to what you are trying to achieve. The information below about this weakening cycle can help you determine which colors / shapes to use sparingly (if at all), based on the elemental nature of your business.

Weakening Cycle

Excess Wood reduces Water = wood takes in water = Wood weakens a Water-type business.

Excess Water reduces Metal = causing rust = Water weakens a Metal-type business.

Excess Metal reduces Earth = earth turns into metal = Metal weakens an Earth-type business.

Excess Earth reduces Fire = earth can put out fire = Earth weakens a Fire-type business.

Excess Fire reduces Wood = burning it = Fire weakens a Wood-type business.

Client Story: Joyce is a client with a printing business that uses primarily paper (making it a Wood element business). Her branding consisted of a completely red background (the Fire element) with white writing. Fire exhausts or burns out the Wood element when used in excess. I asked her how her business was going and she said they had plenty of work, though they never had enough time and the pressure was extreme.

Are your brand colors exhausting your business?

Destructive Cycle

Wood consumes Earth = plants take nutrients = Wood controls an Earth-type business.

Earth dams Water = earth stops water from flowing = Earth controls a Water-type business.

Water extinguishes Fire = add water to contain a fire = Water controls a Fire-type business.

Fire melts Metal = heated to high temperatures, metal melts = Fire controls a Metal-type business.

Metal cuts Wood = metal tools cut down trees = Metal controls a Wood-type business.

Client Story: Geoff is a jeweler who makes jewelry from precious metals. When he suggested using red for his brand, I immediately suggested that it was best he did not. The Fire element of red will destroy the Metal element of his business, melting its success. This business would be better supported with any shade of beige, orange, or yellow and white, or metallic highlights. A hint of blue would also be fine as long as it was not the dominant color for the brand.

Have you chosen a color that will work against your business success?

Conflicting Elements

How you combine elemental colors can also have a big impact on your client's psyche. The elements **Fire (red)** and **Water (blue)** on their own don't really mix well. The startling contrast actually delivers a message of conflict. That's not a feeling you want for your customers as they decide if they will do business with you!

A great example of this is Virgin Airlines Australia, who originally used the colors red and blue in their brand, as well as for the interior of their aircrafts. What a welcome difference it has made since they have changed this to red and purple. How fresh and powerful does this new color combination look and feel?

Here are a few more examples of elemental color combinations that don't work well on their own:

- Yellow and black (add white and metallics)
- Orange and blue (add white and metallics)
- White and green (add shades of blue or black)
- Green and yellow (add shades of red, pink, or purple)
- Red and metallics or white (add shades of yellow or orange)
- Blue and red (add shades of green or brown)

You will note there is a suggested third color to add to these combinations. This will help to harmonize the other two colors. In essence, this takes the elemental conflict out of the combination by adding a bridging color.

Elemental colors that work well together on their own include:

- Red and yellow or orange (*Fire* supports *Earth*)
- Yellow or beige and metallic colors (*Earth* supports *Metal*)
- White and blue (*Metal* supports *Water*)
- Metallics and blue or black (*Metal* supports *Water*)
- Blue and green or (tree) brown (*Water* supports *Wood*)
- Green or (tree) brown and red (*Wood* supports *Fire*)

These combinations work specifically well together in a balanced amount or when the first color listed is made more dominant. For example, yellow (or shades of) and a metallic color used evenly, or yellow being more dominant with a smaller amount of metallics, is optimal. The same applies to each combination listed.

While elemental color combinations might seem complex, there are a few basic concepts you can remember. It is not that you can't use any colors together but rather it's about:

- **How you use them**
- **In what combinations**
- **What amounts of each color you use**
- **The impact you are wanting to achieve by doing so**

Also, take into account all the other elemental factors of your business brand as a whole. The energy of your business type, the words, shapes, pictures (including images of people, hair color / clothing / accessories), and the symbols you choose are all part of a complete concept and the ultimate goal for your brand design is that it is intentional, harmonic, and attractive to your market.

This same knowledge can be used in all your business graphics, and in some cases you might choose to 'break the rules' to your advantage. For example, you might use conflicting colors to draw instant attention or share a serious message with your market, such as yellow and black to indicate danger. While this is not ideal for your logo design and business brand, with other campaigns it might make complete sense.

This is why being conscious of the underlying energy of your design is so powerful. There are so many hidden advantages to having this greater awareness.

Feeling a bit confused? That's OK. Branding your business using elemental concepts is an art. Often it is the slightest addition of a color or shift in the shade of an existing one that can make all the difference. Let a professional help you, otherwise use this final step to know you have the best brand design for your business.

THE RIGHT BRAND FOR YOUR BUSINESS

Once you have taken the time to analyze and articulate each individual facet of your logo and brand design, it is time to pass it on to your graphic designer and have them prepare you a collection of drafts. Pick a designer who is prepared to work with your Feng Shui interests and give you a few samples to play with. Also, be sure they are flexible about any changes you might like to make.

When you receive the samples look at them as a whole design and note your first impression. Check off all the Feng Shui guidelines you have learned throughout this chapter, and then step out of your thinking mind and into your emotional mind. You can do this by asking yourself some questions like: How do I feel about these designs? On a gut level, are these samples right for my business? If nothing feels right don't be afraid to ask your designer to try again.

If you find it difficult to 'hear' your feelings over the 'internal chatter' of your thinking you might benefit from doing a small quietening process first. Close your eyes and breathe deeply for a few moments, tune in to how your physical body is feeling, then ask again:

- **Which sample am I most drawn to?**
- **What do I love about this design?**
- **Is there anything about this design that I don't love, or is just not sitting right with me?**

Once you have come to some sort of conclusion, or even if you haven't, it is a good idea to walk away and do something completely different; disconnect from the project. When you come back you can take a fresh look and often the answer will become clear.

The right design for your business is not something you ultimately decide on an intellectual level; it is something you will know by feel. The Feng Shui guidelines you have learned are not intended for you to overthink your design and end up in more confusion; they are a pathway to tap into your own inner guide. The key to knowing you have created the best brand for you and your business is when you can answer a resounding 'yes' to this question: **Am I completely in love with this design and extremely excited by the prospects of sharing it?** This is your intuition speaking.

Business intuition can take time to develop and your logo and design is a very good place to start. After all, it is not a new concept that listening to your intuition is one of the greatest business tools you can master. Successful business folk have been quoting this for centuries as the ultimate reason they made some of the most triumphant and lucrative business decisions in all of history. They listened to their gut, followed their instincts, or just had a good feel about it. This is the result of refining your 'conscious feel,' or tuning in to your body's vibration, and then as Steve Jobs would say, "... having the courage to follow your heart."

ACTION STEP #4: DEFINE YOUR BRAND MESSAGE

Explore the core of your Brand Message.

If you are starting out in business, use this exercise to assist you in developing your brand. If you are already in business, check in with your existing brand.

1. What kind of business are you in? State the service you offer your clients.

2. Describe the true value of what you offer.

3. Decide what is unique about it. (What makes it different from other businesses of your kind?)

4. Explore how you live your 'Definite Purpose' through this service.

5. Who is your ideal customer? The market you intend to target.

6. What is their greatest desire or challenge you intend to solve?

7. How do you intend to solve this?

Restate your Business Vision.

Now, you are ready to find the perfect colors for your business brand.

1. Select the three colors that you think best resonate with your brand message and explain why you chose them.

2. What shapes or symbols do you think you will use for your business brand and logo? And why?

3. Tick which 'elemental type' best describes your business.

   ☐ **Earth:** grounding, centering, supportive, securing, informing, nurturing, harmonizing, analyzing, nourishing, resourceful.

   ☐ **Metal:** systemizing, focused, clear, planning, structuring, organizing, orderly, refining, completing, hygienic.

   ☐ **Water:** empowering, thoughtful, flowing, serene, enlightening, trustworthy, devoted, faithful, authentic, loyal.

   ☐ **Wood:** growing, developing, healthy, healing, fresh, forward-moving, uplifting, raising, sustaining, achieving, goal-orientated.

   ☐ **Fire:** action-orientated, expressive, expansive, energetic, confident, vibrant, enthusiastic, illuminating, inspiring, passionate.

4. Now, double check for any colors, shapes, or symbols that you believe could be in conflict or hidden 'destroyers' with your business type. If you find anything, what adjustments do you think you could make?

____

____

____

In the next chapter, you are going to give your business brand and marketing material a complete Feng Shui makeover.

Chapter 5
CREATING YOUR BUSINESS BRAND VIBRATION

How to Prepare Your "Feng Shui Framework" for BIG Business Growth

In Chapter 2, we explored the idea that you communicate with customers via the 'vibration' of your business. Remember we said *that as human beings we are drawn to energy that resonates with the highest level of vibration within ourselves, because each of us is designed to grow and expand into our highest possible level of spiritual evolution.* Your ideal business branding will call out to the highest potential in your customers. That's why branding that has the right vibration (energy) unfailingly draws the perfect clients to your business. They are coming to you for the right reasons.

YOU ONLY GET ONE CHANCE ...

You probably finished that sentence in your mind with, "... To Make a First Impression." It may sound trite, but it's true. There are a lot of different statistics about first impressions.

Some say that it only takes seven seconds for a person to form a subconscious opinion about whatever they are exposed to. Malcolm Gladwell, author of *Blink,* says that's way too generous an estimate. He posits that we formulate judgments intuitively and instantly, in the blink of an eye.

We certainly gather information quickly enough. Depending on which study you believe, our physical senses (especially vision) pick up between two and eleven million 'bits' of information each second.

That's why you need to be very careful and deliberate about what energy you project with your branding. Your customers will take in and process this information almost instantly. Not only that, but the first impression and any associated judgments are stored in the customer's memory in relation to your business. So, whenever they see your brand again, these associations will immediately be accessed and reinforced. If you make a poor first impression it can take many, many repeat encounters with your 'improved' brand before the original, negative information is overwritten in your customer's mind.

So, what aspects of your business branding need to be reviewed for vibrational impact? The answer is simple: every single detail. This may sound daunting at first, especially if you think you don't have any kind of roadmap to follow. Fortunately, Feng Shui offers the framework for designing a business brand that makes a positive impression. It also makes the whole creative process more fun!

DESIGN ELEMENTS EXPLAINED

Just as colors and shapes evoke feelings, so do all your other design elements. This includes:

- **The orientation of the design**
- **The quality of the material used**
- **The type, size, and direction of the font**
- **The symbols and graphics that are part of your logo**
- **The wording of any written message**
- **Your photo**
- **The location and layout of each of these elements**

The Feng Shui design principles you will learn in this chapter apply to any rectangular space. Use these concepts in your business cards, webpages, marketing materials, and training products (such as PowerPoints).

Orientation Tips

With any design created in a rectangular space, you have a decision to make about how to orient this shape.

My Business is
Stable and Secure
Trust Me!

A horizontal rectangle represents a good, stable structure; a solid foundation. This is an excellent layout if you are just starting up your business, or you run a business where you want your clients to feel safe with your services, e.g.: medical, accounting, construction industry. It resonates with a feeling of security.

A vertical orientation resonates with a feeling of growth. If you have a strong foundation and are really ready to take off in your business, consider this layout. Also, if your focus is on powerful growth for your clients in some area of their lives, you can take that into account. Remember that this rectangle is the elemental shape for wood, making it ideal for businesses that have this as their core element.

My Business
is Ready for
Exponential
Growth

Join Me!

Selecting Materials

The physical feel and texture of your business card can be very revealing. Most people can tell the difference between cheap and quality cardstock. Choose a material that reflects your business purpose. For example, if you are a clothing designer you could pick a cardstock that is textured to feel like a woven surface. A sustainability or 'green' construction consultant might pick a high-end recycled material.

For brochures, investing in quality paper instead of flimsy copy paper can make a big difference in how your clients feel. A glossy brochure says, "I am excited about my service and want to present it in a way that will catch **your** attention." Cheap paper says, "I can't invest in each potential customer and make them feel individually special."

Fonts. So Many Possibilities!

One of the most challenging design decisions is which fonts to use. There are so many to choose from that you may end up picking something just because it looks cool or unusual. However, it could end up sending the wrong message. Let's look at a few different font designs and what they are saying.

**This font is bold and classically simple.
I am a confident professional. I
know what I am doing.**

*This font is fluid and energetic.
I am a creative thinker. Come to me
when you need solutions.*

This font keeps curling back in on itself.
I lack direction and have difficulty making decisions.

**This font is difficult to read clearly.
I don't communicate well; I might be hiding something.**

Combining fonts can give your message more depth. For example, you could have the no-nonsense, professional part of your business represented with a bold, structured font and the more creative aspect of what you offer clients in a fluid script.

Just don't get carried away and use too many different styles in one design. This looks chaotic. Instead, vary the size of your fonts and use italics and bolding if you want to add more impact to a particular aspect of your business. Here's a good example:

My Business Name

Service I Provide

Tagline or motto

Direction is another area where people sometimes run into trouble. Vertical text is more difficult to read than horizontal text. That's one reason to restrict it to simple words such as short acronyms.

Changing direction and using mirroring is also problematic (unless this message is intended.) This can indicate you are hiding something and it creates a problem with clarity. Your business card needs to be readable at a glance; clients shouldn't have to decipher it like a riddle.

Client Story: Melanie is a graphic artist who uses my Feng Shui services to assist her in crafting her designs. She was hired to design the catalogue covers for a relatively new network marketing company. The company was thrilled with her design draft and was happy to go ahead with it.

Fortunately, Melanie brought the design to me before it went to print. While it was both innovative and catchy, the title on the front was mirrored like in the above example. It made it difficult to read, but that was not the major concern. Networking Companies are already faced with the challenge of being mistrusted and this was only going to reinforce that negative message.

Color is Crucial

By now, you have realized just how powerful color really is. Go back and review Chapter 4 to get a fresh feel for what each color means and the power of the different combinations.

Be aware of using colored fonts on websites and promotional materials as some hues are easier to read than others and some combinations are tiring on the eyes. Too many colors can become confusing and distract the customer from actually doing business with you. It makes good sense to have all your links and sales buttons one consistent color so your customers always know how they can move to the next step with you.

White spaces are known to give the reader a rest and encourage them to stay with your message for longer. This is important to remember with your website design. For your business cards it doesn't hurt to have a white back with just some print, especially if the front is filled with color and information.

Also, bear in mind that you don't have to use the brightest shade of every color. Choose muted colors if you want a more relaxed, calming feel in your design. Use the more vivid colors to create excitement.

Grief Counseling

Party Planning!

What 'Shape' is Your Business In?

When using basic geometric shapes, draw from what you have learned about the five elements in Chapter 4. Remember that each shape relates to an element and a set of qualities. By selecting these shapes as the basis of your designs, you will evoke a similar emotional response from your current and prospective clients.

Let's quickly revisit these qualities and the emotions they evoke:

SQUARE: grounded, centered, supported, secure, nurtured, harmonious, informed.

CIRCLE: ordered, structured, organized, focused, clear, refined, complete.

WAVY LINES: serene, flowing, empowered, enlightened, trusting, devoted, true.

RECTANGLE: fresh, growing, achieving, healthy, forward-moving, uplifted, capable.

TRIANGLE: expressive, expansive, energetic, confident, enthusiastic, inspired, passionate.

Many designs are formed with one of these shapes as their foundation. For example, think about the iconic logo that represents the MasterCard brand, two simple overlapping circles in red and gold. Or the Microsoft symbol which uses a variation of a basic square for their branding.

Another way the elements can be used to empower your brand is by turning your words into shapes. For example, if you are in the business of manufacturing products to use and maintain spas and in-ground pools, the Water element is an obvious choice. For this reason, you might make your text wavy (as long as it still easy to read).

Just remember to always check that you are not using elements that are in conflict with the message you are trying to send or the elemental type of your business, otherwise it may have an adverse effect. For example, using wavy text inaccurately represents unreliability in your business.

Symbols and Graphics

The symbols / graphics / images you choose to represent your business are very potent. Select them with care. Stay away from sharp corners, pointed arrows, or angry looking symbols; anything that feels aggressive. Also, avoid lines that go up and down. That gives an impression of instability and fluctuation.

There are all sorts of symbols that have worked their way into the popular imagination. The dollar sign is a favorite for businesses that assist clients in creating wealth.

However, even positive symbols can make clients feel uncertain about your business in the wrong context. For example, an investment broker would be best not to use a lucky charm symbol like a four-leaf clover. This gives clients the impression that the broker is gambling with their money and counting on sheer luck to make things work out! This symbol would exude the wrong message for most businesses.

Client Story: A local hardware store from an inland country town uses a glorious beach scene for their brand. This aligns with their store name which also has a beach reference. I understand the motivation behind this choice as it represents the owner's big goal; to eventually afford and move their shop to a beachside location. However, I have also witnessed the reaction it gets from prospective clients and even employees.

Firstly, the design is irrelevant to the industry and the local area they operate from. The type of products they sell don't align with a beach concept either. Sending a message of relaxing on a beach is not what a prospective client is looking for in their local hardware store, nor the motivation you would want to project when trying to get them to buy from you. This distracts them as they are bemused by what the logo might mean. However, what is really interesting is how the employees respond.

As the owner handles the sales and management duties and has employees who actually work in the shop, the employees often imply that they do the hard work so the owner can kick back and relax. Obviously this is not the case for anyone running a business; however, what an interesting perspective their employees have arrived at because of this dissonance between the branding and the reality of the business!

Be careful your business branding and graphic designs do not send mixed or confusing messages to your market as well as your staff.

Animal Images

Many people feel a special connection with animals and like to use them as symbols to represent their business. Just like with colors and elements, each animal represents a different set of qualities. Birds are a symbol of communication. They might be a good choice for a person who facilitates communication between parties. This could be a translator or a mediator. Of course, communication takes two! A bird symbol is best when it appears in pairs or groups.

The butterfly is another powerful image. It stands for freedom. A substance abuse counselor might choose this symbol to represent the service she offers clients, support in becoming free from addiction. *We will look at additional animals and what they stand for in a later chapter.*

Alisse's Example: One of my favorite symbols is the lotus flower. It represents the soul and its gift to you is truth. The roots of the lotus reach down into the mud, the stem grows up through the water, and the flower basks in the light of the sun. This represents the connection and progression of our soul from the material earth, through the waters of emotion (vibration), into the glorious sunlight of enlightenment. This message resonates with who I am, what I love to do most and how I wish to be of service to humanity. I have used it on a number of occasions in my business for this reason.

Maybe you have a favorite symbol and story that is relevant to your business.

IT IS WHAT YOU SAY AND HOW YOU SAY IT

The words you use in your business are vital. Words carry a very strong energy. They are impactful and emotive and need to be chosen with care. Use words in business wisely when choosing your name and tagline and when preparing your marketing message and sales copy. Be deliberate about sending an uplifting message. Try thinking about the most important concept of your business and how you could communicate this. Craft a signature phrase that captures the essence of what you offer your clients (your solution). Let's say you are a cosmetics specialist. You might close every email with, "Filling your days with beauty and joy." How do you think that makes your client feel?

Accentuate the Positive

In your marketing it is always a good idea to stay on the positive side. Far too much sales copy is fear-based *(buy our product **or else** you will be overweight, ugly, poor, unhealthy, unloved, unworthy, unsuccessful, etc.).* This approach is based on creating a problem (often imaginary), then the marketing pitch goes on to make clients feel such an overwhelming sense of lack that they 'can't live without' whatever product is being sold. I am not saying this technique doesn't ever work, because it often does (for all the wrong reasons), it just doesn't create the long term experience you would want your clients to ultimately feel.

Use your wording to focus on solutions and opportunities. Talk about the positive and represent your business and your abilities accurately. This will ensure that you attract clients who will actually benefit from your services.

It is perfectly OK to show that you understand the challenges your clients are facing. However, harping on 'problems' just to build up feelings of despair and fear is counterproductive. You are creating negative energy for your business by doing this. Remember, when you make people feel bad, they will eventually decide to spend their money elsewhere.

It Really is All About You

Think about the last time you struggled through a phone system that didn't let you speak to a human being. Didn't that make you feel alienated and frustrated? People like to do business with other **people**. That's one reason to let your clients see your face, no matter how far away they are.

Use a professional photo of yourself on your business cards and your website. In most instances you are your business, and people need to feel connected to you. You don't have to be glamorous or stunning, just well groomed. It is your body language that people respond to. This can come across in a photograph just like it does in person.

Note: If you don't want to be the face of your business, or it doesn't suit your business strategy, then it is vital that you find a face to represent your business. For example, GoDaddy always has a celebrity face on display supporting their brand. You need, at least, to have an image of the type of person that would be the face of your business so it will resonate with your clients. Cosmetic companies may use the face of a woman in the age range of their target customer demographic.

What Your Photo Says

First, your smile tells your client that you are open and approachable. They will receive a warm, courteous reception when they contact you. A forced smile is detectable in a photo because the muscles in your face move differently when you are 'faking it.' When you sit to have your photo taken, smile naturally while thinking about something that makes you happy and excited.

Be sure your photo is a true representation of who you are now. You don't want to finally meet people and have them confused about whether it is really you. This means glamour photos are not a great idea, nor are photos from 10 years ago. If you have changed something about your looks (like your hair cut and color or dress size), have fresh photos done. This adds authenticity to your business. Misrepresenting yourself leads clients to wonder what are you hiding or not comfortable with. It is important not to distract your clients by making them try to 'figure you out.'

If you want to convey a sense of being in charge and able to get things done, stand up. To appear more relaxed, stay seated for your photo shoot. In either case, keep your shoulders back and your posture tall and confident.

Often, you may wish to be photographed at an angle (it is slimming, after all)! Just don't break eye contact with the camera. Your photo needs to show you looking directly out at the viewer. If you are gazing off into the distance or turning away, this makes people feel that you are uninterested or unavailable to do business with. Looking over your shoulder is another no-no as it implies you are too confident and not interested in new business.

Be aware of what you are doing with your arms and hands. Crossed arms can imply defensiveness and not being open. Hands near your chin give the impression of not being sure about your ideas. Pointed fingers are seen as a form of attack.

When it comes to placing your photo within your design, the angle of your body makes a difference. Always have your body facing toward the center of the design. This shows that you support and believe in the message you are sending out to clients.

See how this works?

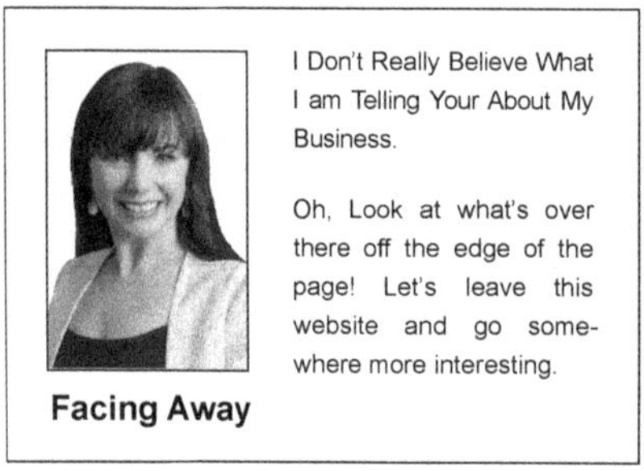

The same principle applies to any image you use in your branding. You don't want the image to be facing away from your information toward the edge of the page. This indicates that the quality you are trying to attract is, in fact, departing.

There are a couple of other mistakes people sometimes make with their photos. Using black and white is one (unless it is in a grayscale text). This is a lovely effect for some situations, but it doesn't suit the image of the modern business. It sends a message of being outdated in your concepts. Better to look 'state-of-the-art' rather than simply artistic.

Plus, when you go with black and white you are missing out on a very important opportunity to communicate. The colors you wear and use to enhance your personal appearance can affect how your clients feel about you. We'll explore how to Feng Shui your wardrobe in a later chapter.

Don't Limit Your Potential

Cropped photos are another area to pay attention to as they affect the energy of the resulting image. For example, if you use a full-length portrait you need to show the whole body. Don't chop off your legs at the knee. This gives the impression of being stuck and unable to move forward.

For a similar reason, don't zoom in too close on your face. This will cut off the top of your head. This indicates that your business will be hindered in its growth.

Alisse's Example: How amazed I was when I walked into a mastermind group in another state and was greeted by a very excited woman. "Oh my gosh!" she said. "It is Alisse Bradley. I am meeting Alisse Bradley in person." I was shocked (and completely flattered) by her words, as I did not know how she knew me.

It turned out she did not exactly know me. She had stumbled upon my website and joined my email list. Both my website and my emails feature the same picture of me. Fortunately, I use a photo that is a fairly accurate representation of what I look like in person. I wouldn't have wanted her to think, "Boy she looks so much older or less attractive in the flesh!" I was really taken aback as to how powerful having my photo on my business brand really was. This woman already felt like she knew me and trusted me. She already had a good feeling about me, and now I was going to get the chance to know her.

Have your clients feel like they know you before they even meet you.

Position and Placement

One of the most fascinating Feng Shui tools is the Bagua Map. This is represented as a square or rectangle divided into nine smaller rectangles. Each 'zone' has a particular area of your business / life it relates to. You can use this diagram to assist you with the layout of your business designs.

Here is a version of the Bagua Map that shows every area of your business

Business Wealth	Business Fame	Business Relationships
Business Health	Chi Center	Business Creativity
Business Belief	Career Path	Travel/Helpful People

As you can see, there is Business Wealth at the upper left corner. Business Relationships are at the top right, etc. Take a look at your current business card and other marketing materials. You may notice that you consistently use some areas and leave other spaces blank.

For example, you might feel very competent when it comes to client relationships. You would place important images or information in that corner. Perhaps you don't feel worthy of wealth and prosperity in your business. In that case, you might find that you have been obviously neglecting that section in your business designs.

This also applies to odd shaped business cards and marketing materials where there are corners cut off. They are not recommended for this reason.

As with all Feng Shui principles, you are best to find a balance between the different aspects of your business. This doesn't mean you need to cram something into every single 'square' of your design. However, you do need to see what business qualities you are consistently missing out on that you would actually like to attract.

Some parts of your design might go especially well in a particular location. For example, your business vision or your tagline would be a good fit for the 'business belief' square. Your photo might feel perfect placed in the upper right-hand corner to stress the relationship aspect of your business. Keep this Bagua Map handy, as you will use it again in other aspects of this book.

POWERFUL PRICING

Feng Shui is a very useful method to use when it comes to pricing your products and services. Numbers play a very big part in the foundational principles of Feng Shui and are in fact the basis of the system used to create the Bagua Map. Feng Shui considers numbers to be tools of greater insight that have been tried and proven true for thousands of years.

In the West, studies have shown that people tend to have very strong (and predictable) subconscious reactions to certain numbers and number combinations. For example, we are hardwired to believe that

$9.99 is much less than $10. Even though we know in our logical minds that the difference is minor, we still perceive $9.99 more favorably. These seem to be deeply embedded reactions that you can't change, so why not use both of these systems to your advantage.

General Tips for Using Numbers Effectively

- Customers are attracted to prices that end in odd numbers. They seem to have a need to complete them. Use the numbers 5, 7, and 9 at the end of your prices to encourage sales.
- The number 8 seems to have this same effect and is the exception to the rule. (There is an explanation for this as you read on.)
- Alternately rounded numbers seem complete and cold, like they don't need us. $100 is not an attractive price, while $97 is.

Magical Number 8

In the East the number eight is considered to be very auspicious (lucky) and relates to a steady increase in wealth and fame. Currently the number eight is also considered 'timely.' We are in the 'age of 8' (from 2004 – 2023). This means eight has even more potential right now. The 'timely' number (8) is said to enhance your current wealth while the number that follows (9) enhances your future wealth. For this reason, prices containing and ending in the number eight and combinations of eight and nine have luck onside and will attract good fortune to your business.

Sales Energy Enhancers

Besides the numbers you use, you can also set up agreements about pricing and sales that have a positive vibration. For example, adding extra value in the forms of bonuses and free products increases the perceived value of what a customer is receiving. Money back guarantees (that you intend to honor to the letter) also raise energy as they build the client's trust in you and your product.

Client Stories: Be aware of how you display your pricing as well, since this can give you clues about your business.

Client A: Tonya is a client who felt like her business was running her life and was on top of her all the time. Clients were ringing day and night and she had no time to herself. When I looked at the layout of her brochure she had all her services and their prices listed up top and her logo with her name and contact details below. I suggested she change them around and place her name at the top and the services and prices below. This symbolized her being on top of her business.

She rang me a week later and said she had found an additional way to solve this problem. She was able to add a second mobile phone to her plan for a minimal cost. This allowed her to use one phone for business alone in business hours only. Just the awareness of what her current layout was saying, and a new decision, helped her set better business boundaries.

Client B: Penny is a client who came to me concerned about her lack of clients. She said that she had plenty of leads but clients often quibbled about the price of her services, believing she was too expensive.

When I looked at her brochure and price list, she had opted to list the prices down the left-hand side and the services down the right. I asked her why and she said it was just something her graphic designer had done for her to be different. I explained to her that by putting the prices down the left-hand side this was the first point of focus for her clients, like she was more interested in their money than pampering them with her service.

We changed this on the design and a month later Penny said her situation was much improved. Clients started to ask questions about the services she offered instead.

See how you can use your design fundamentals to identify inner challenges, and overcome them with outer changes in your business graphics. This is just another way you can use Feng Shui as a success tool for your business growth.

ACTION STEP #5: CHECK YOUR BUSINESS BRAND

Take a look at each feature of your business brand and marketing material and tick off those instructions that your design and message already meets. Ask these questions:

- ☐ Does the combination of my business name and tagline describe who I am, what I do, and what I offer?
- ☐ Do I use my branding consistently for business cards, brochures, banners, and other marketing material?
- ☐ Does my logo design represent my business accurately?
- ☐ Have I double-checked the hidden meaning of any symbols or images I use?
- ☐ Is the orientation of my design (vertical or horizontal) an expression of where my business is and where I desire it to be?
- ☐ Are my business cards, brochures, and other marketing materials of good quality that feel luxurious to the touch?
- ☐ Are the fonts I use easy to read and a good match for my message?
- ☐ Am I professionally dressed in my business portrait in colors that support my branding?
- ☐ Do I appear engaged or distant in my photo? Would **I** want to do business with me?
- ☐ Am I taking advantage of the Bagua Map to enhance the layout of my card to support my business vision? Or, are there blank spaces in the areas that represent the most important parts of my business?

- ☐ Am I playing on customers' fears with my message, or am I offering them an expansive and fulfilling opportunity?
- ☐ Is my pricing in line with the true value of my service or product?
- ☐ Are my prices instinctively attractive to customers?

If you have a website, here are some additional questions to ask:

- ☐ Is my website easy to navigate?
- ☐ Does my website display the same branding as my other materials?
- ☐ Is my contact information easy to find?
- ☐ Are all my links and sales buttons all the same color so my clients know how to do business with me?
- ☐ Do pages load quickly or are they bogged down with too much clutter? (Using too many high-resolution photos, flash, and video can make pages slow to load for some customers.)
- ☐ Is my site hosted on a server that has little or no downtime so my site is always available to clients?
- ☐ Do all of my business emails end with a signature line that makes my clients feel good?

Now that you have your business brand pointed in the right direction, let's take a look closer to home. It's time to Feng Shui your office so it enhances the vibration of your business. Remember, even if clients never see this space, they will sense the affect it has on **YOU**.

Chapter 6

CREATING THE ULTIMATE ABUNDANT OFFICE SPACE

Your Power Position – Element Forms – and Prioritizing Your "Red Phoenix"

The flow of energy in your office space is of special significance in business Feng Shui. That's because constant exposure to a particular vibration reinforces your habits and beliefs on a subconscious level. If you spend 40-50+ hours a week in a particular space, it really rubs off on you. As a business owner you have an advantage that the average worker doesn't. You have an enormous amount of control over how your workspace is arranged and decorated. You can make lots of decisions that will bring beneficial Chi directly into your office. Let's start at the outside of your office and work our way in.

WHAT IS YOUR RED PHOENIX?

The 'Red Phoenix' (as a concept rather than just a specific symbol) is anything that is at the front of the building where you operate your business. That could be your home or a separate office location. It includes the front door, what is beside the front door, the front garden, the boundary fence, the front path, etc. It also includes the front entrance and front window if you have a shop front, as well as the business signage. Even your website's homepage is another form of entryway into your business.

Your 'Red Phoenix' represents opportunities that come to your business (money, recognition, clients, success). How inviting, accessible, functional, and beautiful these things are represents how much success you will attract to your business. Make your 'Red Phoenix' your first priority for office Feng Shui. If you can't entice the energy to your business in the first place, improvements inside the office will have less effect.

The 'Red Phoenix' concept applies to your office desk as well. Make sure your desk is inviting and represents the energy you desire to attract. Be sure it is relevant and beautiful and that there is space in front of your desk for opportunity to flow to you.

Energy Flow in Your Office

Once energy flows in your front entrance, it will be directed throughout your office space. You can determine which way it flows in your workspace by:

- Looking at a floor plan
- Noticing where you naturally want to walk
- Using dowsing rods (if you're familiar with this intuitive practice)

The best flow meanders around the workspace rather than running quickly in long straight lines. You can take advantage of the natural flow of energy by placing products or information you want noticed in the path of the natural flow. For example, you can emphasize the value of service in your business by placing the help desk in the line of natural flow.

You can also direct the flow with furniture, images, colors, etc. Ultimately, clients will appreciate the deliberate flow, as they will feel secure. They can concentrate on your products and services rather than on figuring out which way they should go.

What You See First Sets the Mood

The first thing that you or the client sees upon walking in the door is vitally important. If the focus of your business is that you offer great value at low prices, it's best to see a big savings sign first. If the focus is luxury, a beautifully designed water feature is ideal. In a hotel or other hospitality setting, comfortable chairs will make clients feel welcome.

Of course, the most ideal thing for your customer to see as they walk in the door (on the wall they look directly at) is your business name and logo.

POSITION YOURSELF FOR SUCCESS

In Chapter 3, you learned a little about how your desk placement can make you feel. On the next page there are some diagrams you will find helpful in understanding this further.

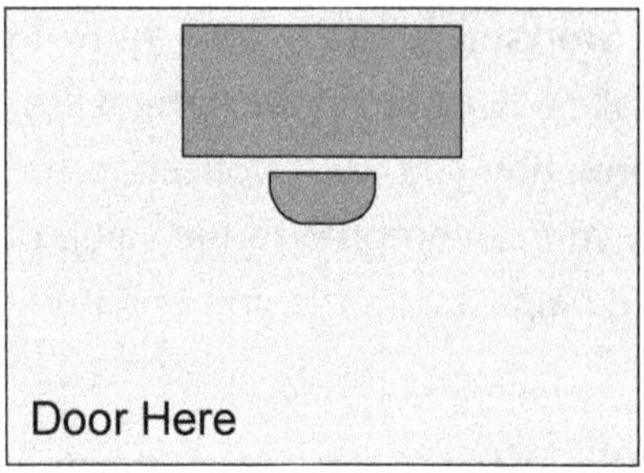

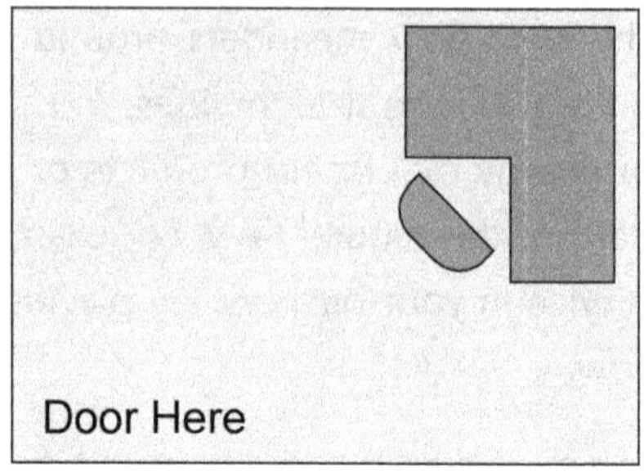

With your back to the door and your face to the wall, you are turning your back to the energy flowing into your office. This is the 'fear' position. With your work surface in a corner, you feel trapped. It will be hard to concentrate in either of these positions. There is, of course, an ideal placement for your desk. Here it is:

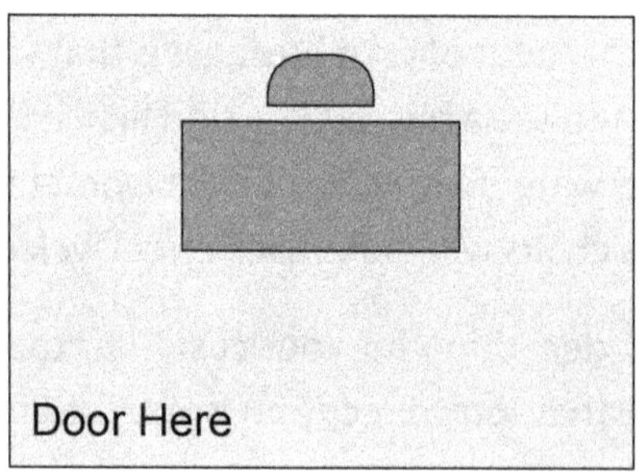

That's your power position. You are facing toward the door. If there is a window in the room, it is good to be able to view it from here as well.

You don't always see this layout recommended in office planning guides. That's because those recommendations focus on 'maximizing' your available space. They want you to tuck your personal workspace into the smallest area possible so you have room for all the other 'stuff' in your office.

Think about what that approach says about your self-worth. Your comfort and sense of empowerment must be your first priority, not worrying about if you have room for another filing cabinet!

Benefits of the Power Position

This layout not only demonstrates your value to yourself. It is also an ideal position for meeting new clients. You look confident and in control of the entire space when you sit facing the door with a wall (strong support) behind you. The layout says, "Welcome to my office. I am a successful businessperson, in charge and completely comfortable in this space. Trust me to help you find the solutions you need."

Of course, Feng Shui is sensed by your clients even if you never meet face-to-face. That's because your office affects you directly every day. Good Chi shines through in your branding and is evident in every interaction with your customers. So, don't neglect your workspace just because no one but you will see it. Remember, it's all about your energy.

Put Your Vision Where You Can See It

In Chapter 3 you learned about the power of images. It is so important to surround yourself with pictures of what you want in your business and life. You definitely must have a powerful success image in front of you at all times. Choose one that symbolizes what you want to be moving toward every day.

This picture will be what you see every time you lift your eyes up from your work. Put a nice big picture on the wall in front of your desk. Use your screensaver as another space to display positive images. Later in this chapter you will find a wonderful list of symbols and their meanings. You can use these as inspiration. In Chapter 10 you'll receive specific instructions on how to create a Goal Board. This will quickly become your favorite item to look at in your office!

Client Story: The pictures you have around you are vitally important for expressing your sense of self-worth. Lisa is a client who had her entire shop space decked out with art that her father had painted. It was a family business. He had now retired and she was the sole owner. However, she was continually challenged with not feeling good enough and she was very focused on her dad's approval all the time. She did not feel like she could succeed on her own merits.

She was an interior designer who had undoubtedly inherited her artistic ability from her father. So, I suggested that she paint her own pictures for her office space and gradually replace her dad's, keeping just one specific wall for her father's pictures in honor of him and her heritage. When I saw her again she was so amazed. From the moment she hung her first picture her confidence started to rise and she is now contracted for her biggest business deal yet.

Here are some more great tips for creating beneficial Chi in your office:

Decorate your office with live, healthy plants and fresh flowers. Discard any dried or drooping flowers. You want to attract the energy of life into your business.

Keep goldfish, really bright gold ones. The best combination is eight goldfish and one black one (the black is to absorb the negative energy in your office). If you keep fewer than nine, choose three or six and make them all gold. These fish represent bringing wealth to your business.

Special Note: Some positions in your office **will not** be ideal for real water and water features. This depends on the individual building and business. Be sure to check with a Feng Shui professional before installing big water features or fish tanks that are difficult to shift.

Incorrect placement may cause financial difficulties. It is better to learn this before you go to all the expense and trouble of installation.

Light your office brightly to bring clarity to your business. Keep your windows clean as well, especially if you are unclear on your future vision. Burn sweet fragrant incense or candles and play uplifting music.

Ventilation is another vital aspect of energy flow since you don't want stagnation in your office. Fresh, moving air is important for productivity and focus. Fresh air equals fresh ideas. The size of the room and the number of people working there will determine the appropriate amount of ventilation. You will need to greatly increase ventilation if there are chemicals being used in the room (printer toner, Wite-Out, glues, paints, cleaning products, cosmetics, deodorants, perfumes, etc.). Clean out the air conditioner regularly. Open windows whenever the weather permits or use a fan to keep air circulating. Just make sure the breeze doesn't create a draft across your desk.

Use colors that resonate with you. Painting the walls is just one way to do this. You can also select furnishings that are colorful. An office chair upholstered in your favorite color might be perfect. Pick a bright, cheerful appointment book and stationery. Keep a selection of pens in different colors, then if you like you can use the one that suits your mood each day.

Alisse's Example: I place an intention on everything in my office. I choose each item very deliberately. For example, I have two owl statues in the entryway to my office. These help me to be wise and always make the best decisions. The elegantly framed image I have of two birds in the southwest corner enhances my ability to communicate with my clients. The giraffe featured on my Goal Board reminds me to stretch for my dreams while staying grounded to my daily tasks. I also love elephants. By placing a bookmark with a picture of them in my daily journal I am reminded to keep the balance, especially between work and home.

I have invested in a quality office chair, which is covered in a gold fabric and features a high back, because I am the queen of my business and this chair is my throne. I have my desk sitting where I can see the door and the window, so I am in the *power position* with my business and my results.

I write with a beautiful pen that was a gift from 'money mindset master' Bob Proctor (featured in the hit movie *The Secret*).[10] It says, 'I am wealthy.' A green leather-bound diary keeps all my appointments in it. This keeps my business growing. My laptop holds all of my business records, so I think of it as my money funnel. It is fuchsia pink which helps me to stay passionate and motivated in my work. It is also the place I go to send messages of love and light to the world. The latest product I have created I call my 'golden egg' (based on the concept of the fairytale *Jack and the Beanstalk* where he finds the goose that continually lays pure gold eggs – this represents continual income for me).

My screensaver rolls over images that represent my goals from each area of my life. Each one has an affirmation that appears on the screen. Because it rolls over it always catches my attention and I have anchored into this a short pause to visualize my goals. This repetition is a very important aspect of creating what you desire.

I have a cream and red mailbox out front with beautiful gold lettering on it. This is where I have anchored that exhilarating feeling of checks coming in the post. Even when there are only bills I pay them little attention and still imagine the excitement of opening a check. I never open my bills at the mailbox, as this is an energy I don't want to associate with this space.

I have a large front door that I make a habit of entering through (even though it is easier to come through the back). I wipe it down often and keep its hinges oiled. I have a designer money frog placed there. This ensures I am always *opening up to* and *attracting in* new customers, plentiful wealth, and boundless opportunities to grow and expand my business.

My car is red so I am always moving forward confidently and energetically. My current key ring has a diamond crystal on it. This makes me feel wealthy when it sparkles in the sun. When I want to accelerate my business success, I clean and service my car. I know this will make everything go faster and travel better.

MORE ABOUT USING COLORS FOR YOUR WORKPLACE

The use of color in your workplace is vitally important. As you have already begun to grasp, color has a greater impact on your business success than you have likely ever imagined. It influences your clients' behavior, appetite, health, purchases, even their whole perception of you and your business. When choosing colors for your office space consider the aspects of your branding, the purpose of the space, the outcome you're looking to achieve, plus the effect it will have on you, as well as your customers. In Chapter 3 we viewed a color chart for your branding. Let's take a more comprehensive look at the energy of color in relation to your workspace.

When considering the effects of color, you can look at it on four levels; physical, intellectual, emotional, and spiritual. In the chart below each color is listed with reference to these.

RED:

Physical: Red is a color that is energizing, strengthening, and revitalizing for the physical body. It improves circulation, digestion, metabolism, adrenal health, and the body's vitality. Its use is beneficial in any space where the objective is to get the physical body moving.

Intellectual: Red enhances the power of the mind where willpower, motivation, ambition, and determination are required. It is beneficial in spaces where leadership development, courage, business / career mindedness, new projects, and results are desired. It stimulates the mind of the pioneer.

Emotional: Red entices feelings of excitement, enthusiasm, passion, and confidence. It evokes feelings of warmth, desire, and wanting to take action. Red is a very powerful color.

Spiritual: Red is of great assistance when the goal is to find and follow your purpose, as well as in the process of materializing your desired outcomes e.g.: money, success, business projects.

Red is best avoided in spaces where concentration, relaxation, and contemplation are required. Too much can increase anger, fatigue, irritation, frustration, and restlessness. Use red strategically.

PINK:

Physical: Pink is a color that works a little like red on the physical body only it is more of a gentle process. It improves the movement throughout the body without over-stimulating it. It increases circulation, immunity, gland function, and joint movement, though at the same time it opens up and relaxes the body.

Intellectual: Pink encourages a positive mindset around aspects such as self-esteem, acceptance of self and others, compassionate thinking, forgiveness, and love; pale pink for self-love, bright pink for Universal love, watermelon pink for 'magic love.'

Emotional: Pink entices feelings of romance, emotional healing, belonging, kindness, femininity, and feeling loved. It feels warm, selfless, comforting, and calming. It is best used in spaces where there is focus on beauty and nurture, relationship development, and the care and compassion of others.

It can assist in spaces where the objective is to heal feelings of loneliness, neglect, trauma, loss, and heartache.

Spiritual: Pink is of great assistance when in pursuit of humanitarian causes, community projects, and when wanting to entice more love into any aspect of life.

PURPLE:

Physical: Purple is a color that is cleansing and calming for the physical body. It assists in reducing irritation and inflammation of the body. It helps to reduce pain, aide sleep, ease infections, ease tension, and reduce hyperactivity and overeating.

Intellectual: Purple encourages the seeking of greater wisdom, thinking outside of the box, expansion, letting go of control, and having trust and faith in the outcome. It enhances mature, reflective, and humane thinking.

Emotional: Purple entices feelings of hope, dignity, faith, presence, as well as belief in self and others.

Spiritual: Purple is very supportive of spiritual pursuits, philosophical development, and greater wisdom. It represents royalty and wealth; often regarded as the holder of higher knowledge.

Be aware that too much purple in a space can decrease the ability to be grounded, make decisions, and take action.

ORANGE:

Physical: Orange is a color that is opening, releasing, and shifting. It enhances appetite in the physical body. It can be used to improve, stimulate, regulate, and boost the body's functioning. Its use is beneficial in any space where the objective is to let go of blocks and move forward.

Intellectual: Orange enhances courage, change, letting go of the old and embracing the new, being open, spontaneity, and having fun. It is beneficial in spaces that encourage play, pleasure, social interaction, belonging, stability, and relationships. It frees and stimulates the creative mind.

Emotional: Orange entices feelings of joy, worth, sensuality, passion, and affection. It evokes feelings of warmth, nurturance, and happiness. It helps with flexibility and overcoming emotional blockages.

Spiritual: Orange enhances deep joy and wisdom. It rejuvenates the soul.

Orange is best avoided where overindulgence is not preferable.

YELLOW:

Physical: Yellow is a color that cleanses, boosts and simulates the body. It improves the digestive, lymphatic, and nervous system.

Intellectual: Yellow enhances learning, logical thinking, clear decisions, concentration, and memory. It expands the mind, is vitalizing, and mentally accelerating. It appeals to the intellect and raises positivity. It is beneficial in spaces where there is a need for study, organization, new thoughts and solutions. It stimulates all faculties of the mind.

Emotional: Yellow entices feelings of enthusiasm, positivity, and being happy. It feels warm.

Spiritual: Yellow raises the spirit.

Reduce yellow in a space if you notice excessive conflict or an inability to relax.

WHITE:

Physical: White is made up of all colors and reflects back pure light. It is hygienic, cleansing, and cooling.

Intellectual: White encourages truth, clarity, structure and completion. It is beneficial in spaces where precision is required, as well as the opening up of new perspectives.

Emotional: White entices feelings of innocence, openness, and taking time out.

Spiritual: White is spiritually protective, cleansing, and expansive.

Excessive white is cold, isolating, and causes loss in direction.

GOLD:

Physical: Gold is a color that improves the physical body's resilience and functioning (longevity). It improves immunity, digestion, and is detoxifying. Its use is beneficial in any space where the objective is to improve the functions of the physical body as a whole.

Intellectual: Gold enhances clarity, optimism, being able to let go of old concepts, improving understanding, and increasing wisdom. It is beneficial in spaces where obstacles need to be overcome, positive change needs to take place, and new wisdom is required.

Emotional: Gold entices feelings of abundance, enthusiasm, and joy. It encourages deep happiness.

Spiritual: Gold is a color that enhances prosperity in all its aspects. It increases higher learning, energy healing, and enlightenment.

SILVER:

Physical: Silver is a color that enhances clarity and consciousness; it helps connect the mind with what is going on in the physical body. It can assist with obsessive disorders where there is a loss of connection with the body and it is seemingly in overdrive.

Intellectual: Silver aids planning, structure, and completion. It enhances the ability to connect with your intuition, decide on your direction, articulate your desired outcome, and make a clear plan to achieve this. It is the architecture color.

Emotional: Silver enhances feelings of magnificence.

Spiritual: Silver assists in the trust of intuition and the ability to verbalize and materialize this.

Too much silver is harsh. A whole feature wall would feel separating; it is best used in smaller amounts and as accents.

GRAY:

Physical: Gray is a color that aids our ability to physically detach.

Intellectual: Gray can encourage independence, self-reliance, self-control, and indifference. Use strategically.

Emotional: Gray is mostly a depleting color. It is melancholic.

Spiritual: Gray is in some cases enticing of time-out, though is not the best color to achieve this.

Gray can feel monotonous, depressive, separate, non-committal, and self-critical. Use carefully in your environment.

BLUE:

Physical: Blue is a color that is harmonizing, sedative, and cooling for the physical body. It has an astringent and antiseptic effect. Its use is beneficial in any space where the objective is to calm the physical body.

Intellectual: Blue is a color that enhances trust, flow, communication, and relaxation. It encourages truth, authenticity, deep power, ambition, commitment, and devotion. It is great to use in spaces where there is a need for trust, openness, honest communication, and calm planning. A turquoise shade encourages being true to yourself and being able to communicate your truth to others. A darker shade indicates authority and power.

Emotional: Blue entices feelings of peace, comfort, nurturance, empathy, support, loyalty, faith, and serenity. It reduces stress and tension.

Spiritual: Blue increases creativity and inspiration. It helps us to connect with our inner purpose and to follow our intended path.

Excessive amounts of blue can make you feel depressed and isolated.

BLACK:

Physical: Black is considered to be void of color. It can enhance relaxation; however, it can also elevate aggression and depression.

Intellectual: Black is a reflective, mysterious, and secretive color. It can be empowering, alluring, and defining.

Emotional: Black can be comforting; however, this is more safely achieved by dimming lights or blocking natural light rather than using as a block color (especially in business). It absorbs others emotions.

Spiritual: Black can be powerful, protective, and enables hiding.

Black is best used with caution. As an accent it can be strong and sophisticated while too much is heavy and tiring. It is safer not to use in large amounts in office décor. Use strategically only.

GREEN:

Physical: Green is a color that balances the physical body. It is harmonizing, healing, calming, cleansing, and relieving. It increases health. It supports the health of the heart. Its use is beneficial in any space where the objective is to get physically replenished or well.

Intellectual: Green is healing for the mind, especially where it has been under a lot of pressure. It improves flexibility, eases the need for control, helps release nonproductive thought, and enhances understanding, acceptance, trust, and faith. Green promotes growth. It is beneficial in spaces where forward movement is desired. A more olive shade encourages patience; an apple green entices new ideas and projects; while a tree green encourages sturdiness and sustainability.

Emotional: Green entices feelings of peace and balance. It is relaxing and calming and restores the body when it has been under stress. Certain shades promote freshness and cleanliness, while others promote sustainability and reliability.

Spiritual: Green assists in knowing the truth, identifying who you really are, how you feel, and in the creation of abundance in your life.

Overuse may be too calming and inhibit action.

BROWN:

Physical: Brown is a color that enhances earthiness. It assists the body to relieve tension.

Intellectual: Brown supports common sense, stable thinking, and the ability to commit and be dedicated.

Emotional: Brown is sensual, soothing, and nurturing.

Spiritual: Brown is grounding. It can be beneficial for business that promotes connecting with nature.

Brown is a color that can be very elegant, especially in the right shade. Too much brown can make you feel stuck.

Yin vs. Yang Colors

All colors come in varying shades. They can be light or dark, warm or cool, muted or bright. Dark, cool, and muted colors are more Yin, which is quieter and more relaxing. Light, warm, and bright colors are more Yang, therefore exciting and stimulating. Dependent on the ratio of one to the other you can create a different influence.

The concept of Yin and Yang is a piece of Eastern philosophy referenced in the *I Ching*[11] (Book of Changes) in approximately 700 BC (however use of this wisdom is evident even further back). In this work, all phenomena are said to be reduced back to Yin and Yang: sun and moon, masculine and feminine, day and night, light and dark. This is also relative to the Universal Law of Gender. All things must contain both components to generate life, and should either become too dominant, the other will eventually weaken and be consumed by its counterpart. Therefore, it is vital to consider some sort of balance in your business environment. This includes the colors you use for your workspace.

Yin Colors

Yin colors make an environment feel cool, moist, passive, restful, introspective, smaller, and slower.

The yin energy relates to the female, the moon, and the night. Yin colors include any that can be described as being dominant in any of these three qualities: dark, cool, and muted. They are colors that 'recede' when you look at them. (Note: If you squint your eyes when you look at your environment you will be able to see which colors recede – move back – and which come forward.) Generally, these are blue, green, and violet-based colors.

Excessive **yin** in an environment will feel cold and dark and tend to lower activity. The occupants in this workspace will be inclined to lack vitality and motivation and could become lazy. This environment will encourage low productivity, procrastination, and will not effectively entice new business.

Yang Colors

Yang colours make an environment feel hot, dry, active, motivated, extroverted, expansive, and fast.

The yang energy relates to the male, the sun, and the day. Yang colors include any that can be described as being dominant in any of these three qualities: light, warm, and bright. They are colors that 'come forward' when you look at them. Generally, these are red, orange, and yellow-based colors.

Excessive **yang** in an environment will feel bright and hot and tend to over-stimulate activity. The occupants in this space will be inclined to be overactive and scattered with little ability to relax. This environment will encourage anxiety, incomplete projects, and business 'coming and going' with little results.

The Right Balance is the Answer

When choosing yin or yang colors for your workspace take into consideration the purpose of the space. A massage therapist would benefit from making the therapy room more **yin** while a gym would benefit from making the exercise room more **yang**. An office space or front reception is always best to be slightly more yang as this will stimulate action and results, as well as make customers feel welcome to do business with you.

There are other aspects of your environment that will influence its yin-yang factor. The types of furniture you have, the fabrics you choose, the decorating features you add, and how much light the room has (both natural and artificial). The general rule is, the heavier and darker the aspect looks and feels the more **yin** it is, while the lighter and brighter the aspect looks and feels, the more **yang**.

Ask yourself these questions. Do you feel like you are lacking energy, motivation, and results in your business? Are you struggling to stimulate customer interest in your product or service? Maybe you need to add more yang to your workspace. Alternately do you feel like you are too anxious and scattered with too many projects on the go? Are clients coming and going but not doing business with you or lacking loyalty? Maybe you have too much yang and need to add a little more yin.

Client Story: Terri is a client who was struggling to stay motivated in her business. She would begin the day with much enthusiasm though once she was in her office she would find herself lacking energy and not achieving the tasks she had set out to do. Terri sold educational products online: her office space was tidy and functional; however, she had not paid a lot of attention to the design elements of her environment as she did not see her clients face-to-face. All the walls were painted in a light gray, the floor and her work desk were a dark heavy timber, the curtains were dark blue in color, and there was not much natural light. While the furnishings were beautiful it became immediately obvious that her environment was very **yin**.

To balance this out Terri painted a bright jade green feature wall behind where she sat at her desk, placed a large white floor rug with jade, magenta, and golden yellow patterns on it under her desk, added a chandelier floor lamp to the far right-hand corner of the room, and hung a colorful print of children playing in a garden on the wall. She also placed a bright yellow vase with three sunflowers in it right beside her computer where she spent the majority of her time. Terri's energy shifted immediately; she now loves being in her office and she finally feels like she is being productive.

Is your office décor helping or hindering you to achieve your to-do list?

Here are some more hints for using color in different work environments:

Example of colors for an office space

Use light blue, orange, and in particular yellow, to maximize concentration and focus. Minimize the use of white, brown, and black, as this will have the opposite effect. Use only highlights of red as this may cause restlessness and eyestrain. This environment would benefit by being slightly more **yang**.

Example of colors for a therapy room

Use purple and violet for spiritual connection. Use blues, greens, and indigo to induce relaxation and healing. Add touches of orange for creativity, red brown for grounding, and small hints of yellow for stimulation. This environment may benefit from being slightly more **yin**, dependent on the outcome desired for the participants. Dimming natural light can also help to achieve this.

Example of colors for an eating area or restaurant

Enhance appetite and socialization with orange. Add greens for harmony and good health. Blue to enhance communication. Where there is an inclination to overeat, add more blue; for under-eating use orange. For stimulation purposes and digestive function this space is best to be more **yang**.

Example of colors for a recreational or sporting facility

Use warm tones such as reds and oranges to increase energy and enthusiasm, greens to improve health. Hints of blue for relaxation spaces and cooling down areas. This environment is best to be more **yang**, to stimulate movement.

Example of colors for an entry or waiting room

Dependent on the type of service you offer, to create a social atmosphere use orange, add green for harmony and balance, pinks for love, and blues for communication.

If your goal is to provide your customers a space to relax prior to seeing you then blues are best, still including the greens and adding hints of yellow, red, or orange for interest. This environment is best to be more **yang**, taking into consideration how long you expect or desire your customers to stay.

USING ELEMENTS TO FENG SHUI YOUR OFFICE

The concept of the elements is also relevant in your office environment. Every facet of your office décor represents one of the five elements either by the material it is made from, the color it is, or the shape it is in. An environment that has good Feng Shui also has a good balance of each of the five elements (in one of its forms), making it a productive space to work from.

The productive cycle that is necessary to sustain our natural world is relatable to the cycle of flow within your business (since everything is connected). You will learn more about this in the final chapter of this book.

Here's how the elements support one another in nature:

PRODUCTIVE CYCLE

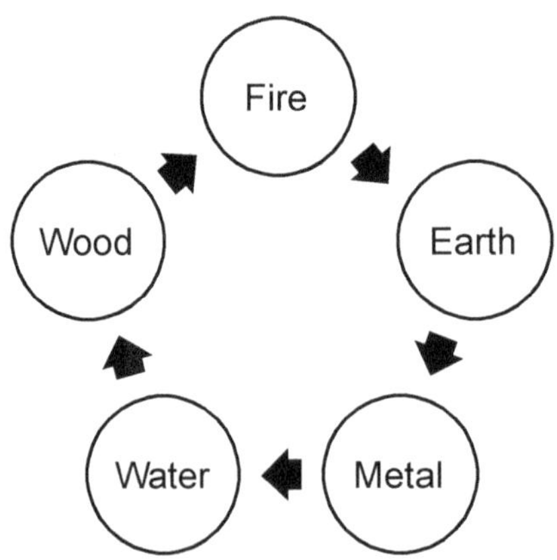

Fire – Earth – Metal – Water – Wood

- **Fire creates Earth** – burned wood becomes earth
- **Earth produces Metal** – metal deposits form deep in the earth
- **Metal condenses Water** – drops of water form on the outside of a can
- **Water feeds Wood** – plants and trees need water
- **Wood fuels Fire** – if you have a low campfire, add more wood

To create a productive elemental flow in your workspace, stand at the doorway and take a visual audit of all the furnishings and décor that are present. Again, take into account wall colors, curtains, prints, equipment, and lighting; everything that you can see. Then affirm that each element is represented in one of these ways. (Also note that no element is overly dominant.)

MATERIAL

Fire – any object that produces real fire, anything electrical, anything plastic.

Earth – anything made from earth, such as pottery, ceramics, crystals, rocks, mud, and carpets.

Metal – anything made from metal.

Water – anything that contains real water, anything glass.

Wood – anything made from wood and live plants.

COLOR

Fire – red, pink, burgundy, purple.

Earth – yellow, orange, earth tones.

Metal – white, gold, bronze, silver, copper.

Water – blue, black.

Wood – green and tree brown.

SHAPE

Fire – triangular or pointy.

Earth – square or flat on top.

Metal – round or dome.

Water – wavy or irregular.

Wood – rectangular or tall.

In Chapter 3, we learned that each element is associated with specific feelings or vibrations. We have also learned in earlier chapters that the way we are naturally drawn to set up our environment can teach us about ourselves. Here is what the elements teach us.

Not enough of a particular element in your environment may reflect or increase certain behaviors you exhibit:

- Not enough Fire – Lack of **Passion.**
- Not enough Earth – Lack of **Awareness.**
- Not enough Metal – Lack of **Planning.**
- Not enough Water – Lack of **Belief.**

- Not enough Wood – Lack of **Growth.**

In reverse, too much of a particular element in your environment will reflect or increase certain feelings:

- Too much Fire – **Burn Out.**
- Too much Earth – **Feeling Stuck.**
- Too much Metal – **No Direction.**
- Too much Water – **Procrastination.**
- Too much Wood – **Frustration.**

Working with balancing the elements is another simple strategy to assist you in identifying some of the weak spots in your business. You can then rectify this by adding or lessening the needed elements. This doesn't mean you won't put other systems into place to assist with these challenges; however, you will be surprised at how quickly the right opportunities begin to present themselves when you shift the elemental energy in your workplace.

Client Story: Even the slightest changes can make a huge difference. Freda is a client who had designed her entire office in black and white. She absolutely loved the effects of the contrast and wanted to keep it this way. When she learned about the Feng Shui elements she was concerned that even though she loved her office it was not considered 'good' Feng Shui. When she took a closer look at the five element lists she realized that in fact, four of them were present. The furniture Freda had chosen was veneer and with the inclusion of all her technological equipment, the Fire element was represented; the floor was a terracotta tile covered with a large woven rug representing the Earth element; the abundance of the color white represented the Metal element; and the black represented the Water element. While black is not considered an ideal color for every office space, what Freda had done well was counterbalanced it with a lot of white (yin-yang balance), with slightly more white featured than black. She was also a photographer for a fashion magazine so the theme aligned well with her.

However, the one element that was notably missing in the environment was Wood. Freda easily rectified this by adding a large vase with an abundant floral arrangement to her décor and three of her best pictures of models wearing floral print garments on her back wall. She also added other hints of green throughout the office; such as a mouse pad and pen holder, a cushion on her chair, and a small valance over the window. After a little convincing she purchased a live plant and potted it to the right of her entry door. Just these small changes were enough to ensure the space had a balance of all five elements, and that Freda's business would not be stunted by lack of growth (which to no surprise was her current business challenge). Freda was very happy with the shift in her office space and grateful for the results it mustered (three new clients the very next day). She says she can really notice the difference.

It just feels so right.

Using Feng Shui Figurines

Figurines are beautiful and fun to use in Feng Shui. Place a money frog (a symbol of wealth) facing inward at the front door of your business to bring money in. Place a turtle at the back door to slow the flow of money out of your business.

Or, you can place three one-dollar coins on an ornamental ship. Put the ship in the upper left corner of your desk (the wealth corner in the Bagua Map). Make sure it is sailing toward you to bring you wealthy clients.

Are you facing a problem you need to solve quickly? Bring out a monkey figurine and put it on your desk. You can even put it on top of the paperwork involved in whatever project you are finding challenging. The monkey is a symbol of problems solved.

Do these little statues have magical powers? Well, whether they do or not, one thing for sure is that your mind has some pretty awesome abilities. These figurines act as visual reminders of the intentions you have 'set your mind' on attracting. That's why it is beneficial to know what each symbol stands for, so you can make **conscious**, deliberate choices about what to create.

At the same time, you will notice that the meaning of the symbol is very aligned with the behavior or the story around its source. For example, below you will read that dogs represent loyalty because, by nature, dogs are loyal.

Every symbol you will learn about in this chapter has its own meaning. Do you need to use them all just to be on the safe side? No, that would indicate a lack of focus; it also gives off a desperate vibe. Just choose the ones that are significant for you and what will complement your décor.

FENG SHUI SYMBOLS EXPLAINED

Angels: The angel is a reminder that you are not alone. Help is at hand; you only have to ask.

Bells / Chimes: These attract positive energy and peace to your office space.

Birds: Communication and good relationships are represented by two or more birds. A flock of birds brings opportunity and continual good news.

Cat: This symbol raises your sense of independence and self-esteem.

Buddha: The fat, laughing Buddha represents all the good things in life!

Butterfly: Use this image when you want to be set free from something in particular. Just use one as multiple butterflies can create discord in relationships.

Candles: These glowing symbols connect your wishes with universal power. Choose colors and scents that resonate with you.

Dog: The faithful dog represents protection and loyalty. This image will remind you to be true to yourself.

Dolphin: The spirited dolphin represents joy and spontaneity. Do you want to make your work fun? Bring a dolphin into your office.

Dragonfly: This insect symbolizes complete transformation. Use it in preparation for a major business change.

Eagle: This visionary, far-seeing bird signifies gathering information to make the best choice in your business. This is a powerful help in long-term planning.

Egg: The egg is a sign of new life. Choose a jade egg crystal to bring good luck to any new project.

Elephants: These wonderful animals offer wisdom, strength, balance, belief, and power. Place them anywhere, as long as there are two or more together and they have their trunks up.

Fu, Lu, and Sou – The Three Star Gods: Fu is for health and family, Lu is for wealth and abundance, and Sou is for longevity. Place these guardians of prosperity in a high position so they can overlook your business.

Frog: This is a wonderful symbol of wealth and abundance. As mentioned before, place the frog facing inward at your door to bring money into your business.

Giraffe: The giraffe is reaching for the sky. Bring this animal into your office to create the right atmosphere when you are stretching for what you want.

Globe of the World: This powerful symbol is your perfect choice for expansion and growth. It is a sign for International Business.

Horse: Use the horse to bring you the energy of action and endurance. It's perfect for those times when you need to work late and still be productive. Make sure the horse is not running out your office door as you may get the opposite reaction.

Jade Plant: This Chinese money plant attracts abundance into your business when you place it at the front door. Take good care of your jade plant and make sure it stays healthy. Green potted plants generally symbolize growth (except for spiky ones like cactus).

Kangaroo: Did you know that a kangaroo cannot jump backwards? This beautiful animal indicates continual forward progress.

Lion: The lion is a symbol of courage. Use this image when you are facing a business challenge and need an extra dose of daring.

Lotus: The lotus flower is a lovely representation of your soul. Place this symbol in your office when you would like to find more purpose in your business.

Monkey: The clever monkey is a natural problem solver. Bring in this animal to improve the health of your business and when you need to find solutions.

Moon: The moon represents intuition and accessing the deeper side of yourself. Use it when you are soul-searching.

Mountain: A mountain provides support for you in your business. Place a picture behind where you sit in your office.

Owl: The owl is a symbol of wisdom and knowledge. Use this sign when you need assistance in making the right decision about a complex issue.

Pagoda: This towering structure indicates an increased level of fame and recognition for your business. Let the pagoda attract the attention your business deserves.

Phoenix: This is another symbol of transformation, like the dragonfly. However, this sign is best used when you want to access the power of the Fire element. Don't mix it with Water signs.

Quan Yin: This is the Female Wish Buddha. She eases business burdens for women and gives clarity. Quan Yin is a symbol of loving compassion. Let her support you.

Rooster: This cocky bird is an image of confidence. Use this symbol to boost your sense of pride in your business accomplishments.

Sunflowers: These bright, cheerful flowers represent golden opportunities. Be prepared when you use these, exciting business prospects are coming your way!

Tortoise / Turtle: This is a symbol of support, security, and strength. Place the tortoise at your back door to prevent your abundance from flowing out.

Client Story: Mercia is a client who really loved the idea of using Feng Shui to improve her business results. Her resistance was she was not drawn to the Feng Shui trinkets she thought were needed to implement this sort of luck. She had a preconceived idea that Feng Shui was only about Chinese Dragons and Fortune Cats sitting at the front register. Mercia's boutique sells designer handbags and her market is women looking for a luxury item. She didn't feel Feng Shui would be a fit for her business.

Mercia was relieved to hear that this is not what I was recommending for her to do. That Feng Shui is about achieving flow, beauty, and abundance in alignment with the design style that is most suited to the business type you operate. Once Mercia understood this she was easily able to implement all the Feng Shui luck she desired using furnishings, colors, and symbols that enhanced her environment. In fact, what she found was her store felt more luxurious; she attracted more of her ideal clients and naturally her sales increased. Her clients commented on how much they enjoyed coming to her boutique and how impressed they were with the improved quality of her products (even though they were exactly the same). How funny.

See how Feng Shui can work its charms in any environment. That is because it is an enhancement tool and not an interior design style in itself. This understanding will be a big help when you start to implement it into your business.

Eliminate Poison Arrows in Your Office

There are a number of things in your office that can make you feel uncomfortable. These are called 'poison arrows.' Having corners pointing directly at you or sitting with a beam running directly over your head are a couple of examples. Shelves that are at head height are another feature you will want to eliminate. These feel threatening (and can give you a nasty bump if you aren't careful). The position where you spend the most time must be free of poison arrows!

- Bookshelves can be problematic if they are facing toward you. Use a book cabinet with doors to conceal the shelves. Or, bring all the spines of the books even with the edge of the shelf to create a more balanced, less jagged effect.
- Your wastebasket needs to be out of your sight. Empty it regularly and don't let any bad smells accumulate in your office.
- Broken, neglected, and offensive items that are facing you are also disturbing. Remove or repair these accordingly.

- Double-check your front entryway for any poison arrows. This would include power poles, corners of other buildings, or traffic that drives straight towards your front door. These could be upsetting the flow of energy into your business.

Final Check-in

Now that you have implemented Feng Shui in your work environment there is one final thing for you to do. It's time to connect with the overall vibration of your space. Just as with your Branding when you used your intuition or 'gut' feeling to check-in with your design, now do the same thing for your workspace.

Stand at the doorway and attune yourself to the space. Take a few deep breaths and begin to observe all of its aspects. Ask yourself, "How does this environment make me feel? Do I like it? Do I feel good in it? Is it the ideal representation of who I am, my business, and the outcome I want for my clients?"

Identify if there is anything in the environment that doesn't *feel* right. Do you think it needs to be moved to improve this? Is there anything that you really don't like or is bothering you? Does this item need to be removed completely? Do you feel like there is anything you need to add to your environment? What do you think this could be?

Make the changes you *feel* are needed and once you have, check-in again until you can confirm you *feel* really good in your new surroundings.

ACTION STEP #6: CHECK LIST: CREATE YOUR ABUNDANT ENVIRONMENT

The Mobile Effect: Imagine a mobile made up of twenty-seven pieces on nine strings. If you were to gently tap just one of those pieces, within seconds the whole mobile would move, like a ripple effect. Feng Shui in your business works exactly the same way – because it works on *energy* – and just like the pieces of the mobile, all energy is connected. This means you don't have to get everything *Feng Shui perfect* for it to have a positive effect on your business; you simply need to implement what you can when you can. Then, over time, you can take further steps towards changes. Same as for the mobile, the more pieces you tap the faster and more abundant the movement; however, know that every progressive step you take will improve all facets of your business in synchronicity. This is a good reason to begin whatever Feng Shui you can today.

If you already have an established workspace you can apply this checklist and do your own Feng Shui assessment. If you are creating a new workspace or revamping one, use this checklist as a guide.

1. Drive by or walk by your workplace and see if you feel drawn to enter it. Is it easy to find and easy to enter? Can you sight any signage? Is it in good condition?

2. Check the front of the business. Is it beautiful and abundant? Is it clear of any blockages and the front door easy to enter?

3. What do you see directly in front of you as you walk in the front door? Is it inviting? Can you see your business logo clearly? If in an office, can you see your business registration on display?

4. Have you checked your business flow? Where do you feel directed to walk? Is there easy flow throughout your office space? Do you need to shift or redesign anything?

5. Is your desk in the power position? Do you have a beautiful chair that is comfortable? What do you see directly in front of you? Does it represent your business goals?

6. Is the space clean? Is the lighting adequate? Do you have good ventilation? Do you have live plants? Are they well cared for? Is the aroma pleasant? Do you have soothing or uplifting sounds?

7. Check all the pictures, images, calendars, and symbols for hidden meaning. Does anything need to be moved or replaced? Do you have a strong image you can see from your desk? Do you have your representations of your skills and qualifications on display?

8. What are the dominant colors in your environment? What do they represent? Does that match how you want you or your clients to feel?

9. Check the yin-yang balance of your environment. Is your environment slightly more yang than yin, unless the alternative is more suitable?

10. Evaluate elements (see below)

Evaluate the elements in your office space. Put a check mark by each of the items below that you find in your office environment.

Note: To decide what element an aspect represents look at its most dominant feature or standout characteristic (material, color, or shape). For example, a red painted wooden cabinet is dominantly red or the *fire element*. Pale green floor tiles are dominantly ceramic or the *earth element*. While a round shaped glass picture frame is dominantly round and the *metal element*. Use your own judgement / intuition to identify this.

EARTH FORMS

- ☐ Anything made from earth such as pottery, ceramics, rocks, mud, crystals, terracotta
- ☐ Carpet or rugs
- ☐ Objects that are square shaped, flat on top
- ☐ Objects that are yellow, orange, or earthy in color

METAL FORMS

- ☐ Anything made from metal such as filing cabinets, bookcases, metal statues, cutlery, coins
- ☐ Clocks that chime with a metallic sound
- ☐ Objects that are round or dome shaped
- ☐ Objects that are white, gold, bronze, silver, or copper in color

WATER FORMS

- ☐ Anything that contains water such as fish tanks or fountains
- ☐ Glass items, mirrors, windows, drinking water glasses
- ☐ Objects that are wavy or irregular in shape
- ☐ Objects that are blue or black in color

WOOD FORMS

- ☐ Anything made of wood such as furniture or books
- ☐ Live plants
- ☐ Tall, rectangular shapes
- ☐ Objects that are green or tree brown in color

FIRE FORMS

- ☐ Any object that produces real fire (fireplace, heater, etc.)
- ☐ Electrical equipment (computer, printer, etc.)
- ☐ Plastic objects
- ☐ Objects that are triangular or pointy shaped
- ☐ Objects that are red, pink, burgundy, or purple in color

See if there is a fairly even balance among all five elements. If not, eliminate one or more items from the most dominant category. If one element is missing entirely, bring in an object that represents that element. Make sure it is an object that you like and that feels prosperous and beautiful to you.

11. Check for poison arrows.

12. Have you stood at the door and done a complete assessment of your environment? Have you asked yourself how you feel about everything? Does this business environment truly represent the interior style you resonate with? Are you completely happy with this space? Does your space reflect how you want your business to be?

Chapter 7

EXPOSURE TO ABUNDANT SENTIMENTS

How to Link Your Business Vibration to Your INFINITE Potential

There are countless ways you can use Feng Shui to increase the good Chi of your office space and your business. In this chapter you'll learn strategies that will boost your prosperity potential even more.

SETTING INTENTIONS WITH FENG SHUI

To begin this chapter, let's look at the power of intentions. In a sense, intentions are just another kind of (deliberate) anchor. They can potentially connect everything in your surroundings with what you desire to attract into your life. There is no limit to the ways you can use positive intention to link everything in your life into the realization of your infinite potential.

One particularly beautiful example of the power of conscious intention to affect our outer reality was found in the work of Dr. Masaru Emoto. His team took photos of distilled water that was frozen to produce ice crystals. There was a striking difference between the before and after photos taken based on whether the water was exposed to a positive or negative influence. Exposure to sentiments such as, "You make me sick. I will kill you," resulted in deformed and chaotic crystal patterns with no symmetry. In fact, water exposed to negative intentions created crystals similar to those that form in polluted water. In contrast, exposure to the prayer of a Buddhist monk, classical music, aromatic oils, or loving intentions resulted in beautifully formed ice crystals.[12]

One experiment Dr. Emoto's team performed that has yielded excellent results involved simply taping messages of gratitude to a bottle of water. This is something you can do at home. Write an affirmation onto your water bottle. Then set the intention that your water will fill you with that supportive message every time you take a sip. Your body is 70% water. It makes sense to nourish your physical being with positive intentions.

- If you desire business growth, you could use a label on your water bottle that says: "I feel the unlimited, expansive energy of the Universe flowing to me and through me."
- If you have problems with insomnia, you could label the water you keep on your nightstand: "My instant sleep potion."
- If you have a bottle of water you take to the gym, you can label it with an intention such as: "I am stronger today than ever before!"

You can set intentions in your food as well. Try: "I am eating this salad to nourish and revitalize my body." That not only creates a healthy vibration around your food, it also makes you really aware of what you are eating. You would probably feel awkward saying, "I am eating this donut to nourish and revitalize my body." Once again, anchoring your intentions helps you change your habits for the better!

There are lots of ways you can use intentional Feng Shui tips in your business. Here are a few to get you started:

Making things beautiful helps anchor in positive feelings. So, buy yourself a lovely pen and anchor into it that: "I am a success," or an expensive pen with the anchor: "I am wealthy."

You can sign your checks by placing a smiley face on the back and anchor in your gratitude for the service that has been provided to you. If you have created a green folder for your bills, paste a picture of someone successful on the outside. That way, when you put your hands on the folder you can anchor in the thought, 'If they can do it, I can too.' That will make you feel good.

Remember this Feng Shui practice when it comes to your customers. Add that special touch by sending correspondence in a beautiful envelope or having an attractive header on email, etc. People are receiving things all the time. If they receive something special from you, they will anchor that thought into how they respond to your business. This makes it well worthwhile to go that extra mile. With the right business branding, you can make everything beautiful while still being professional.

10 WAYS TO FENG SHUI YOUR SOUL

Are you finding it challenging to set positive intentions that you believe in, deep down? That may be because you are awash in a culture of negativity. You'll find it much easier to set intentions and feel their positive influence if you cleanse and renew your own energy. Here are 10 great ways to do this. Try them and see how much happier and peaceful you feel!

1. Go on a media diet. Bad news sells; so it is rare that the daily newspaper or online news updates have something positive and uplifting to share. Stop watching and reading about all this turmoil and strife. It impacts your thinking / feelings and lowers your vibration. If you wish to enjoy some entertainment, choose a positive movie, a feel-good story, or a book that focuses on what you want to attract into your life. How about something that raises your awareness of how to live a more empowering life?

   If you hear something often enough you will believe it. This is how we develop our beliefs in the first place. We hear something once and we are not sure if it is the truth. We hear it a couple more times from different sources and we begin to expect that it is the truth. We hear it several more times and we begin to believe it. After hearing it enough times, we have complete faith that it is the truth. Build the belief-set you *desire* to have and use it to replace the one instilled in you by a negative, fear-based culture.

2. Be aware of the radio station you listen to and the lyrics of the songs you hear. Often the lyrics of music can be negative, focusing on tragedy and loss or violence. This can be particularly destructive when we are not paying attention. When that happens, we do not make the choice to consciously reject the information, but rather absorb the messages straight into our unconscious mind. (Remember the unconscious mind is the part of us that does not reject information, it accepts everything that is presented to it.)

   Talk radio shows that are all about arguing and fighting over politics or social issues are designed to trigger emotions of anger, hate, and fear in listeners. People actually get addicted to this strong emotional response, but it is very counterproductive to personal growth. Replace negative radio programming with positive and uplifting stories of people who have achieved much in their lives.

Listen to inspiring music (soothing or joyful instrumental music is great if you aren't in the habit of paying attention to lyrics). You may even choose to create your own collection of songs with empowering lyrics only. This is easy to do with a smartphone, an iPod, or MP3 player. It can come in very handy as a vibration raising tool first thing in the morning or before important meetings.

3. Always create a Goal Board of pictures of those things you want to attract into your life. Really open up your opportunities by including your big goals. Using Feng Shui and the Bagua Map, you can position your pictures to be more powerful. By knowing your own personal energy number you can be even more precise about placement. (We will talk more about the Bagua Map, plus finding your energy number and success direction, later in this chapter.)

4. Realize that everything you have in your environment is a 3D goal. Really think about this. Everything from your task chair to your wastebasket to your office desk to the colors you have chosen in your décor, all of these are 3D Goal Boards which are influencing the way you act and attract. Be sure everything in your office reflects your soul's desire.

5. Associate with people who live the type of life you want to live. If you hang around with successful people long enough, you will become successful because you will learn their ways. We all learn via association. This is because our physical senses have the ability to pick up information from our environment at an astonishing rate. Again, most of this information is absorbed directly by our subconscious minds, where our beliefs are set and stored.

6. Study the lives of exceptional people, especially those who are leaders in your industry. Listen to their audios, attend their seminars, read their books. Then integrate what they do and have done to become successful. You can even model the way they dress, their habits, the way they hold themselves, their websites, their methods of business, etc. (as long as they are aligned with you). Be sure to find people who are truly living the life you desire to live, in all its aspects.

7. Practice what you preach. If you intend to ask others to buy your product or hire your service, then you must be prepared to do the same yourself. Use what you sell and instill what you teach. A true leader is someone who has *knowledge* combined with *experience*. There is something very magnificent about someone who walks their talk. People can read this on an energetic level.

8. Create space in your environment for expansion. If you need to de-clutter then do this now! If you have already de-cluttered and you are still crammed in, you may need to invest in a bigger filing cabinet or office space. Have a diary with big enough pages to write all of your appointments plus your future goals and gratitude statements. By making space we have the power to attract new and different things into our life. If your environment is already full, you are saying, "I do not have space to grow."

9. Cleanse and renew the energy (Chi) in your space. The energy in your office can become dense, especially if you deal with a lot of clients or have a challenging day. This energy can be cleared by opening your windows, even for just 10 minutes at the end of each day. You can also do this by using essential oils such as lavender. Be sure that your office smells fresh and pleasant.

10. Invest in gifts and fresh flowers for yourself and have them on display in your office. This is a constant reminder that you value who you are. Treat yourself to a beautiful office chair. Make sure

it is comfortable, has a high supportive back, and makes you feel like you are sitting on a throne. Remember you are the king or queen of your business ... honor yourself!

Alisse's Example: When I first committed to these 10 Soul Steps I did not fully understand the profound impact they would have. To begin, it wasn't easy to adopt these new habits. My first awareness was how engulfed I was by the outside world. It is not a wonder that change is so challenging. Though, once I tuned out all the extraneous 'noise' in my life, I was able to hear myself.

As an entrepreneur I was stuck on the business treadmill, chasing every latest trend or new concept, trying to keep up with my competitors as well as outrun the doom and gloom of a world in crisis. I rarely gave myself any time to connect with what I believed was my truth. (Note: If you really pay attention to the media, the world has been in crisis in one form or another throughout all of time, and we buy it!)

As a result, I experienced overwhelming stress and anxiety. Add to that poor health, low self-esteem, and being overweight and out of shape, as I had absolutely no time for me.

Now I know, while the media is someone's truth, it is not necessarily mine. I make a deliberate choice as to what I allow into my energetic space and I only let in what is going to elevate me towards what I want to achieve in my life.

It goes something like this: you get really quiet, you listen to your soul, you pick your perfect target, aim, and fire from your deepest truth. There are no hit and misses anymore.

This has meant I work fewer hours to achieve the exact result I am looking for, allowing me to live a balanced and joy-filled life. **I encourage you to try this for yourself.**

EVEN MORE FENG SHUI FOR YOUR OFFICE

Whew, do you feel better now? Great! Let's look at additional Feng Shui strategies for your office space to make it even more energetically inviting and stimulating as you create business success.

Office Building Selection

- If you intend to lease a building to use as your office, find out the history of the building and the history of the other businesses that have been there. What were their reasons for leaving the office space? If they have outgrown it and had to get a bigger space, this is a good sign. If every business that has been in there has closed down or moved quickly, beware of the energy of that space.
- It is best not to choose an office space opposite a cemetery, hospital, school, large power station, etc. These are all poison arrows and project negative energy towards your building.
- Look at the road the office is on and how it makes you feel as you drive up to your business. Does it reflect what you are trying to create or not? Do you feel good as you drive toward it? Is it easy to park your vehicle and access the office?
- Make sure you have adequate signage on the front of a business or an office. It needs to be the right size for the business and in proportion with the size of the building or office space. It needs to be clearly readable. The design of this sign is vital.
- Be sure that the front door of your business or office can be seen. If it is easy for people to walk on by without noticing, it is easy for the positive energy to pass on by and not enter either.

- Whenever you have a water fountain outside of a business or a home, make sure it is shooting upwards or flowing towards the premises. This will indicate money coming into the building. (Reminder: Be sure to consult with a Feng Shui Designer before installing a water fountain / feature. Its position is critical for your wealth.)
- Avoid an office with a blue roof. This will make you feel like you are financially drained all the time.

Office Layout Design

- Have your business logo immediately visible as you enter your office. Shine a bright light on it to raise the strength of your business.
- If the front door is aligned with the back door, position a plant somewhere in between to slow down the flow of energy.
- In a plan layout for a house or an office it is never good Feng Shui to have a bathroom or a toilet in the center. This will cause drainage of good energy.
- A square or rectangular-shaped room is most auspicious for business.
- As mentioned earlier, get conscious about your office flow. Buildings have flow and they direct clients in certain ways. Maybe you have a product to sell and you want to direct a client to the product table. You need to stand at your own front door and feel where you are motivated to walk. This is the flow of your workplace.
- Also, you need to feel the flow from the outside of your workplace and where you feel directed to walk. If you share an office space with other businesses and the flow is better directed

to their office, this may slow down your success. Deliberate placement of plants and signs can change this flow.

- Always place your desk in the power position with a solid wall behind you and your face to the door / window if possible. (This tip is vitally important, which you have likely gathered by now.)
- Make sure you can open windows into your office. If this is not possible then live plants are a must for this space.
- Always enter your workplace through the front door. Throughout history, only the paid help entered through the back door, not honored guests or the owner.
- Don't have too many glass windows in an office. If you are working with too much light behind you, you may be missing opportunities. Plus, with a glass window at your back you may not feel supported. Use a blind to remedy this and put it behind where you sit. If you are speaking in a function room with windows behind you, close the curtains before you speak.
- Be sure the area in front of your office desk is clean and free of obstacles. It is best if you can take two clear steps out in front of your desk.
- If you have a long corridor to your office this may make it difficult to get new clients. If you have a long corridor within your office space, with three doors in a row, shut the middle door to keep your finances on track.

Office Accessory Selection and Placement

- What is your first impression when you look at your office or enter your front door? Is it tidy, do you have clutter? Is your desk messy or organized? What pictures do you have on your wall? Can you find everything you want the moment you want it? Do you always feel energetic and motivated in your office, or tired and lacking in enthusiasm?

- Check the pictures on your calendars and ensure they are abundant. This is an important item. When you look at a calendar, you associate it with your time and future events. Abundant pictures include a sunrise for new beginnings, sunflowers for golden opportunities, flowers for all that blossoms. Avoid images of stormy weather or a dry barren desert.
- Your office keys represent what you think and feel about your work. Make sure your keys have nothing on them that is broken, nothing that has no purpose (such as keys you don't use), and nothing you do not like. Put all of your keys on one ring as opposed to linking one ring to another and then to your keys. Face them all in the same direction. Place them on a key ring that makes you feel good, positive, abundant, or powerful. House keys are also important, especially when your office is at home. If you are continually losing your keys you may be having challenges with knowing your direction. Tidy up the key ring and add a symbol such as a dragonfly.
- Do not sit at the corner of a table when in a boardroom; this is a poison arrow aimed at you. Choose a seat along the sides. If you need to be in control of the meeting, then sit at the end of the table that is facing the door.
- Make a significant change in your workplace at least every three years. For instance, a new piece of furniture, fresh paint on a wall, or change of a featured wall picture.
- Remember that your chair is your throne; it needs a high back and armrests for full, comfortable support. Also, have enough

room to swivel your chair all the way around so you do not feel limited in any way.

- When using antique / second-hand furniture try and learn its origin. All furniture holds past energy and so it needs to complement your business vision. Antique furniture creates a feeling of stability; too much feels resistant to change and inhibits business growth.
- Avoid sitting at the end of a long corridor, as the energy that flows towards you is too aggressive. This will make you feel under attack.
- If you are working all day on your computer it is a good idea to place crystals or a salt lamp beside your computer to correct any negative energy.
- Do not place your frogs and money symbols over a drain such as a sink or a toilet. This represents sending money down the drain or toilet. In fact, don't place anything in your bathroom that you do not want to flush down the drain or toilet; this includes any luck symbols, water creatures, photos of friends, positive sayings, or calendars (that would represent flushing time down the toilet). Bathroom and toilet doors need to be kept closed. Also make sure you always put the toilet lid down. Toilets are best left clear of decoration; they should simply be clean and fresh. Never place diplomas on walls that adjoin to bathrooms.
- Be careful of abstract paintings. These can feel like confusion and make your clients unsure of your ability.
- Excellent lighting at your entry is important.

- Red is an auspicious (lucky) color to use for business. It attracts business recognition. Introduce some red to your office space, but do not overdo it.
- Avoid glass furniture. This type of item gives no support.
- Have a good-sized office desk. The larger the desk the more opportunity and business expansion you will attract. A rectangle shape is best. Be sure it is not too big for the room.
- Remember the rule for mountain pictures: they need to be behind you rather than in front of you so you are supported and not blocked. Mountains with three peaks ensure you will be of great influence in your industry. Three separate pictures behind you will have the same mountain effect (even if they are not of mountains).
- Ring a bell or use a singing bowl through your office. This awakens new energies into the environment.
- Crystals energize the office and enhance creativity; small crystal balls are an excellent choice and can look great placed strategically.

Animal Symbols

- The stork and crane represent long life. Use these birds to represent the longevity of your business. A picture of 100 birds in front of your desk can attract opportunity to your business.
- If you like dragons be sure not to place them near each other. This creates conflict.
- For a more productive day, start work in the hour of the dragon, between 7am and 9am.

- A rooster placed in your office will remedy any negative gossip that is harming you in business.

Other Symbols

- The pagoda is a symbol of knowledge, fame, and career. The more levels it has, the more powerful it is. Place it on an office desk to enrich a career or on a bookcase to enhance learning.
- A picture of a compass will help you find the ideal direction and make the best decisions for your business. Place the picture of the compass on your bank ledger, invoice book, or diary. This way, answers will continue to be presented to you.
- A ship with gold coins (not a warship) coming towards you will bring wealthy clients to your business. Just be aware of overusing wealth symbols. You do not want to feel desperate since that sets a negative goal.

Office Plants

- It is a good idea to have plants at the entry to your workplace. These represent growth, abundance, and the conversion of negative energy to positive. Plants can also be placed in front of a poison arrow, such as a corner directed at a desk. Have plants with round, abundant leaves, not spiky or cactus-type plants. Bonsai trees are not good Feng Shui as they are stunted and represent no growth in your business. Bamboo plants and peace lilies are excellent. The pine tree represents being true to yourself. Place a jade plant in your office where your clients can see it to enhance prosperity (not at the back door). The African violet represents peace and harmony.

- Fresh flowers are always good Chi for an office. This ensures fresh business will come your way. Keep the water clean and replace them as soon as they begin to wither.

Client Story: It always amazes and delights me how well these Feng Shui strategies work. Susan is a client who desired to begin a home-based business in massage. Her challenge was that she worked full time and needed the income to sustain her life. I suggested that she make her entry more appealing: place a plant on either side of the door, lay down a red rug, and hang pictures which represented what she wanted in front of her as she entered. She loved to garden, so she actually planted some pretty flowers along the sides of the front path. She called me two weeks later to say that, after 20 years of working in her full-time job, she had been offered a part-time job *for the same amount of money*. This was going to allow her to start her business in the spare hours. Even better, her part-time job was a position in a health food store, which connected her with clients who were already interested in a natural approach to health and wellness. She was very excited.

Start making the changes you can today, and watch new doorways open up!

THE BUSINESS BAGUA MAP AND COMPASS DIRECTIONS

The tortoise is a very special symbol in Feng Shui, as it is at the core creation of the Business Bagua Map you first read about in Chapter 5. Legend has it that in approximately 1950 BC Emperor Yu, the founder of China's first dynasty Xia, observed a giant tortoise emerging from the River Luo. On its back was a grid of nine numbers that were represented by dots. The dots were in a specific sequence whereby no matter which direction they were added up – horizontally, vertically, or diagonally – they always added up to the number 15. This is what is referred to as the 'magic square' and it looks like this.

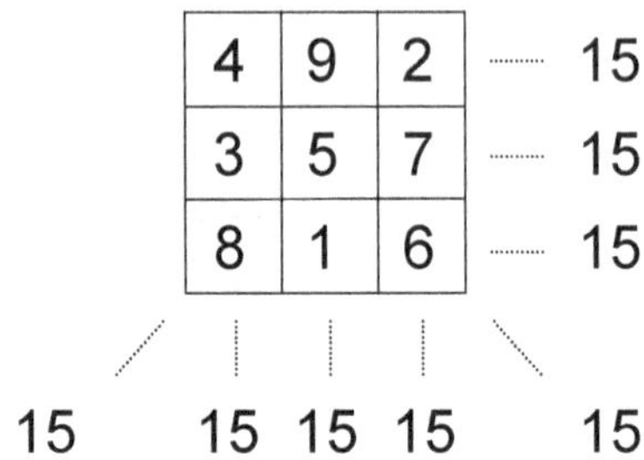

This magic square is the template for the Business Bagua Map. It is considered a superior form of ancient wisdom as it is created from numerology (the foundational language of the entire universe). In Chapter 5 you learned that this map could be used for your business branding to assist you with layout. It can also be used over your workspace to identify which sectors relate to the different areas of your business. There are a few methods as to how this map can be applied. In this book you will learn the Compass System, as it is considered to be the most accurate for general workspace use.

Articulating the Compass directions of your work environment is a great tool to identify which areas represent the different parts of your business. You can then activate these using furnishings, colors, images, symbols, etc. in alignment with the basic design of your business environment. To do this we first need to revisit the map and learn its compass directions and the elements that apply to each sector. This information has been added to the following diagram.

(South-East) **Business Wealth** ***Element is Wood***	(South) **Business Fame** ***Element is Fire***	(South-West) **Business Relationships** ***Element is Earth***
(East) **Business Health** ***Element is Wood***	(Room Center) **Chi Center** ***Element is Earth***	(West) **Business Creativity** ***Element is Metal***
(North-East) **Business Belief** ***Element is Earth***	(North) **Career Path** ***Element is Water***	(North-West) **Travel/Helpful People** ***Element is Metal***

Once you have identified the compass directions you will need to know the direction of each corner of your workroom. If your business has more than one room do this for each of them individually.

The Bagua Map has nine sectors, the middle sector being the Chi Center and representing the center of the room. The other eight sectors will rotate depending on the compass directions you identify. Imagine laying the grid over the room matching the sectors to your compass readings.

If you have an original plan of your workplace or home office and can locate the north symbol you may be able to devise the different directions this way. If not, draw yourself a mud map, divide it into nine sectors, and follow this simple exercise. Stand in the middle of the room with your compass (if you don't have an actual compass most smartphones have an app for this) and facing each corner, record what direction it is. Then write down the business sector and element this aligns with from the Business Bagua Map provided. For example, if when you face the far left corner of your room the compass reads west, then write down that this sector is your Business Creativity corner and the ruling element is Metal. Once you have identified each facing direction and the Business area it represents, you can then activate the area from the chart below. Use this information as a complementary guide to support the style, flow, and beauty of your existing environment.

Here are some examples: If the corner to your immediate right when you walk into your consulting room is in the southwest, then this is your Business Relationships corner and is ruled by the element of Earth. If it makes sense, you might place two yellow chairs with a small veneer coffee table there (Fire element supports Earth) to enhance relationships in your business. If the area is not suited, then perhaps an image on the wall of two people enjoying each other's company would work better. Avoid placing a single chair or a picture of a single woman with nobody to talk to. This would detract from the energy of this corner.

If the far right corner is in the southeast, then this is your Business Wealth corner and is ruled by the Wood element. If the wall space is suited you might put an image in a timber frame of three gold fish swimming; if this looked awkward or wasn't harmonious with the interior design of your office then you might choose to feature a model sailing ship or a blue glass vase with a silk floral arrangement. If none of these suggestions 'worked' then you could simply place three gold coins in a drawer in this area. You would avoid having an overfilled filling tray, a broken lamp, or a stagnant fish tank in this space.

The most important factor in both of these examples would be to design the space with the intention of enhancing this area of your business. The below chart is a guide to what materials, shapes, colors, and images activate each of the Bagua sectors. This does not mean you fill your whole office with mismatched furniture (unless this is the look you are after) or need to include every element listed. It is a guide for activating these areas to assist you in creating what you want in your business.

Note: The evolution of the Bagua Map directions and elements applied are based on centuries of observation of the earth, the skies, the animals, the plants, as well as human inhabitants. Their movement, changes, patterns, influences, as well as repeated consistent outcomes. It encompasses the most advanced mathematical and scientific knowledge (as do all of the ancient systems / philosophies included in this book). Meticulous comparison, analysis, and synthesis of the phenomenal behavior of all existence has turned this information into the simple list you are about to read. An expansive insight into not only what is obvious but far beyond what the eye can see is why it works so powerfully and predictably. It truly is a fascinating and exciting method for those who want to apply it.

ACTIVATING THE BUSINESS BAGUA MAP

Chi Center:

Ruling Number: 5

Compass Direction: room center

Ruling Element: Earth – pottery, ceramics, crystals, rocks, mud and carpet, shapes that are square or flat on top, yellows, oranges, and earthy tones.

Complementary Element: Fire – real fire, anything electrical, plastic, triangular or pointy, reds, pinks, burgundy, and purples.

Destroying Element: Wood – real wood and live plants, rectangular or tall shapes, greens and browns.

Enhance this area by adding Earth or Fire elements.

Images and Symbols: this space is the center of the room. It can be left clear, or include a simple floor rug. In a smaller office the desk might be situated here. This will depend on the room layout and purpose of the space. Again, remember flow is most important.

Business Wealth:

Ruling Number: 4

Compass Direction: southeast

Season: spring

Ruling Element: Wood – real wood and live plants, rectangular or tall shapes, greens and browns.

Complementary Element: Water – real water and glass, wavy and irregular shapes, blues and black.

Destroying Element: Metal – real metal, round or dome shapes, white, gray, or metallic.

Enhance this area by adding a Wood or Water element.

Images and Symbols: any that represent money, wealth or abundance, a ship sailing, a water symbol, a lush plant, sunflowers, real money (coins are an exception to the metal rule here).

Business Fame:

Ruling Number: 9

Compass Direction: south

Season: summer

Ruling Element: Fire – real fire, anything electrical, plastic, triangular or pointy, reds, pinks, burgundy, and purples.

Complementary Element: Wood – real wood and live plants, rectangular or tall shapes, greens and browns.

Destroying Element: Water – real water and glass, wavy and irregular shapes, blues and black.

Enhance this area by adding a Fire or Wood element.

Images and Symbols: any that represents what business fame looks like to you; articles of recognition, a lush plant, a candle or a lamp, an image of the sun (not setting), images of entertainers, birds flying.

Business Relationships:

Ruling Number: 2

Compass Direction: southwest

Season: late summer

Ruling Element: Earth – pottery, ceramics, crystals, rocks, mud and carpet, shapes that are square or flat on top, yellows, oranges, and earthy tones.

Complementary Element: Fire – real fire, anything electrical, plastic, triangular or pointy, reds, pinks, burgundy, and purples.

Destroying Element: Wood – real wood and live plants, rectangular or tall shapes, greens and browns.

Enhance this area by adding Earth or Fire elements.

Images and Symbols: anything that represents harmonious work relationships or partnerships, a representation of love, a pair of something, an image of the earth, a crystal ball, something feminine.

Business Creativity:

Ruling Number: 7

Compass Direction: west

Season: autumn

Ruling Element: Metal – real metal, round or dome shapes, white, gray, or metallic.

Complementary Element: Earth – pottery, ceramics, crystals, rocks, mud and carpet, shapes that are square or flat on top, yellows, oranges, and earthy tones.

Destroying Element: Fire – real fire, anything electrical, plastic, triangular or pointy, reds, pinks, burgundy, and purples.

Enhance this area by adding Metal or Earth elements.

Images and Symbols: anything that represents creativity, music, fun, joy or life, technological symbols, creative projects, a picture of a lake, beautiful artwork.

Travel / Helpful People:

Ruling Number: 6

Compass Direction: northwest

Season: autumn

Ruling Element: Metal – real metal, round or dome shapes, white, gray, or metallic.

Complementary Element: Earth – pottery, ceramics, crystals, rocks, mud and carpet, shapes that are square or flat on top, yellows, oranges, and earthy tones.

Destroying Element: Fire – real fire, anything electrical, plastic, triangular or pointy, reds, pinks, burgundy, and purples.

Enhance this area by adding Metal or Earth elements.

Images and Symbols: anything that represents business expansion and travel as well as business support or mentorship, your address book or business cards you have collected, a symbol of a tortoise, artwork that features human hands, astronomy, a metal feature, a male symbol.

Career Path:

Ruling Number: 1

Compass Direction: north

Season: mid-winter

Ruling Element: Water – real water and glass, wavy and irregular shapes, blues and black.

Complementary Element: Metal – real metal, round or dome shapes, white, gray, or metallic.

Destroying Element: Earth – pottery, ceramics, crystals, rocks, mud and carpet, shapes that are square or flat on top, yellows, oranges, and earthy tones.

Enhance this area by adding Water or Metal elements.

Images and Symbols: anything that represents where you would like to see your business direction go, business logo, water symbols, image of a pathway, landscapes containing a river or creek, the moon, a clock, something that moves.

Business Belief:

Ruling Number: 8

Compass Direction: northeast

Season: late summer

Ruling Element: Earth – pottery, ceramics, crystals, rocks, mud and carpet, shapes that are square or flat on top, yellows, oranges, and earthy tones.

Complementary Element: Fire – real fire, anything electrical, plastic, triangular or pointy, reds, pinks, burgundy, and purples.

Destroying Element: Wood – real wood and live plants, rectangular or tall shapes, greens and browns.

Enhance this area by adding Earth or Fire elements.

Images and Symbols: anything that represents business or personal beliefs, spiritual symbols, images of wise people (Buddha / Einstein), books, business certificates in frames, an elephant symbol, picture of a mountain (as long as it is not in front of you), candle or a lamp.

Business Health:

Ruling Number: 3

Compass Direction: east

Season: spring

Ruling Element: Wood – real wood and live plants, rectangular or tall shapes, greens and browns.

Complementary Element: Water – real water and glass, wavy and irregular shapes, blues and black.

Destroying Element: Fire – real fire, anything electrical, plastic, triangular or pointy, reds, pinks, burgundy, and purples.

Enhance this area by adding a Wood or Water element.

Images and Symbols: any that represent business health, musical instruments, image of a tree or forest, a lush plant, a recipe book, a national emblem, a horse or monkey symbol, antique items, family picture.

In Chapter 10 of this book you will revisit the Bagua Map to assist you in creating your Goal Board. You will find more lists of images relevant to each sector there. These can also be applied to your workspace where relevant.

Client Stories: Here are some examples of how clients applied the Bagua system in their business.

Example A: Margaret is a client who produces a magazine for women who want to build successful businesses. She struggled with staff turnover, never seeming to be able to find the right people to do the work she wanted. Often she found herself working excessive hours having to do or re-do most of the tasks herself.

In the northwest corner of her office (Travel and Helpful People) she had a veneer bookshelf cluttered with paperwork, research material, and equipment she used in her business. She didn't feel good about this and had been thinking to make some changes. Using the Bagua Chart, I suggested she replace this cabinet with a metal one that had sliding doors on it. She was still able to place all the same items in the cabinet, though now it was all behind closed doors. Across the top of the cabinet Margaret placed a rich yellow table runner and a statue of the Archangel Uriel. Margaret had an affinity with the angels and for her Uriel represented expansion and help from the Universe; it was also a stunning statue. On the same cabinet I suggested she place an engraved metal box and put in it a note of exactly who the helpful people she needed for her business were and a picture of the world globe. On the other side of her angel I had her place a cover of her most recent magazine in a gold frame.

Not long after, Margaret remembered an old colleague she worked with in her last job. She took a chance that this lady might be seeking a career change and organized to have coffee with her. It turned out she did, and now Margaret employs her as a project manager and has her source the exact people the magazine needs.

Example B: Derek is a qualified builder who was working for a large building company. He had his license to do renovations under his own banner; however, he could not generate enough business to do this full time. He had a belief that there was not enough work out there and too many other builders doing the same thing.

Derek had not paid much attention to his home office as his actual work was done on job sites; it was just quoting and paperwork he did at home. When we applied the Bagua Map to Derek's office we noticed that the entry was in the northeast corner, so my recommendation was that he place framed copies of his qualification, builder's license, and an award he had won on the immediate left wall as he walked in. He had a beautiful pair of floor-sized elephants he purchased for the living area of his house that I suggested he also place along this wall looking inwards. They sat nicely just in from the door without blocking the walkway. Derek had built himself a desk from some beautiful old timber he found on a job site; however, it was over in the far right corner facing the wall. We located the Fame area (south) that was at the center back of the office, turned the desk around, and placed it in this sector. He purchased a quality green leather chair with a high back to give him the feeling of being the *king* of his business. On the back wall behind him I suggested Derek take photographs of three of his favorite renovations, enlarge them, and hang them in timber frames. The concept of three pictures matched the three peaks of the mountain you read about earlier in this book. On his far left desk corner we placed a lush pot plant and a business card holder with his newly printed cards in it, even though clients did not visit his home office.

What was interesting was that clients did start to visit his home office and Derek felt confident to have them there. Over the next 12 months Derek was able to fully fill his schedule with his own work and has been in business for himself ever since.

Example C: Careen was a client who had been at home raising her children for over 10 years. Before she started her family she had worked as an accountant; however, she had become very interested in effective parenting and been pursuing a degree in Counseling from home. Careen was keen to begin a part-time business since her children were now older, though she was very torn as to what her actual area of interest was. She was very unsure of which business she wanted to pursue.

Careen's office had lovely flow and design already, as she was very creative. Using the Bagua Map we were able to locate the Career Path (north) area of her space and use it intentionally to help her with her decision. On my recommendation she put her creativity to work and painted a beautiful picture of the moon in shades of blue and silver. She framed it in a silver frame and placed it on the northern wall. This would assist her to access her deeper unconscious and connect with what she truly wanted to offer in her business.

Interestingly, Careen discovered that she did want to work with parents and teach them effective strategies to raise their children; she added art therapy to her qualifications and enhanced her counseling with the use of painting and color to help her clients connect with and express themselves. She also found that her accountancy background was of huge benefit as one of the big areas parents struggled with was budgeting and knowing their financial boundaries with their children. Careen now runs a successful counseling business that she works in and around her family and is achieving huge results with her clients.

Now you can use the Bagua Map to activate the different areas of your business and create the outcomes you are looking for. Being even more specifically attuned to your environment gives you another huge advantage in creating and attracting more growth, success, and sustainability in your business.

ACTION STEP #7: DISCOVER YOUR SUCCESS DIRECTION

Find Your Success Direction

As promised, here's the secret to finding your energy number and your personal success direction.

> **Note:** While the compass directions are specific to your business environment, your success direction is individualized to your birth date (Heaven's Luck).

The formula (for women) to find your Ming Gua or energy number is as follows: Add the digits in your birth year until they become a single digit. (For example, if you were born in 1978, this would be: 1+9+7+8=25. Then, add again: 2+5=7. The single digit is 7). Next, add 4 to this number: 4+7=11. Again, reduce the number to a single digit: 1+1=2. The number **2** is your energy number.

For men, the formula is to add the digits in your birth year until they become a single digit (as above 1978 reduces to 7). Then, subtract this number from the number 11 (11-7= 4). Add the numbers again until they become a single number (we don't need to in this example). The number **4** is your energy number.

However, it's just a little more complicated than that in two ways.

First, in the Ming Gua system the number 5 is not considered 'real' as a personal energy number. For women who end up with the number 5 in the equation above, use the number 8 as your lucky number instead. For men, substitute the number 2 for the number 5.

Second, the Chinese year system used to find your energy number doesn't match up with the calendar used in the Western world. It may run from late January to February, from February to January, or from February to February. Sound confusing? Don't worry. Here's a chart showing 22 years worth of dates and corresponding energy numbers.

Year of Your Birth	Dates From – To	Female Number	Male Number
1960-61	28 Jan-14 Feb	2	4
1961-62	15 Feb-4 Feb	3	3
1962-63	5 Feb-24 Jan	4	2
1963-64	25 Jan-12 Feb	5 (8)	1
1964-65	13 Feb-1 Feb	6	9
1965-66	2 Feb-20 Jan	7	8
1966-67	21 Jan-8 Feb	8	7
1967-68	9 Feb-29 Jan	9	6
1968-69	30 Jan-16 Feb	1	5 (2)
1969-70	17 Feb-5 Feb	2	4
1970-71	6 Feb-26 Jan	3	3
1971-72	27 Jan-14 Feb	4	2
1972-73	15 Feb-2 Feb	5 (8)	1
1973-74	3 Feb-22 Jan	6	9
1974-75	23 Jan-10 Feb	7	8
1975-76	11 Feb-30 Jan	8	7
1976-77	31 Jan-17 Feb	9	6
1977-78	18 Feb-6 Feb	1	5 (2)
1978-79	7 Feb-27 Jan	2	4
1979-80	28 Jan-15 Feb	3	3
1980-81	16 Feb-4 Feb	4	2
1981-82	5 Feb-24 Jan	5 (8)	1

Now that you have your personal energy number, you can figure out the best direction to energize your office with your business success goals.

SUCCESS DIRECTION BY YOUR PERSONALIZED NUMBER:

1	**= southeast**
2 / 5 (men)	**= northeast**
3	**= south**
4	**= north**
6	**= west**
7	**= northwest**
8 / 5 (women)	**= southwest**
9	**= east**

What to do with this information:

1. Make sure the corner of your office that is located in your success direction is beautiful, as well as relevant to your business goals. Use the Bagua information provided to activate this area. Treat this corner as your specific space to activate your own personal energy.
2. Avoid keeping clutter, broken or unused items, stagnant water, anything that is dying or dead (dried flower arrangements), items you don't like, incomplete projects, or your rubbish bin in your success corner.

3. Write your business vision on a goal card and place it in this corner. You can slip it into a draw or frame it if you wish. Place it with three gold coins in a red envelope, with a citrine crystal, or with your favorite luck symbol if you like.
4. Once you create your Goal Board, you can also place it in your success corner if that works well with your office layout.
5. If you are able to face your success direction when sitting at your desk, or place your desk in the success sector, without compromising the 'power position,' then you will be able to maximize on your most fortunate energy.

Alisse's Example: Here is what I have written on my Goal Card and placed in my southwest corner. My personal Ming Gua adds to up to the number 5, so I work with the energy number **8**.

I, Alisse have true wealth.

I calmly allow money to flow into my account, over $X (exact amount written) each week.

I easily accept sales and easily accept that my soul is worth all the abundance I earn.

I have balance; I relax and allow myself to enjoy my business and my family.

That or something better...

Dated: XX/XX/20XX **Signed:** Alisse Bradley

PART THREE

Chapter 8
MATERIALIZING SUCCESS WITH YOUR INTERNAL MINDSCAPE

The Truth About Affirmations – Visualization – and Greater Intentions

By now, your office is likely to look and feel pretty good. So, let's take another peek within your internal 'mindscape' to help you move to the next level. Feng Shui is a fantastic tool to create business abundance. Now it is time to take a step up and discover how to materialize anything you desire as we learn the truth about affirmations, visualization, and greater intentions.

This is the greatest truth of all: you can have anything you want right now if you can master these tools of the trade.

AFFIRMATIONS – TOOL OR TROUBLE?

Positive affirmations are a very popular tool for materializing your reality. However, they need to be used correctly. Neal D. Walsch, author of *Conversations with God*, gives a great example when he suggests not starting a statement with "I want," such as, "I want to be rich." This indicates that you are in a state of 'want.' When you keep saying you *want to be rich*, you keep creating a reality of *wanting to be rich.*[13]

Your affirmations are also better when not stated as beliefs about what may happen in the future (e.g. "I believe I will one day be a successful businesswoman."). Affirmations aren't pleas you are sending out into the ether hoping someone hears them. They are actually directed inward; you are developing a new set of beliefs for your subconscious mind. The repetition is designed to train your mind in a new way of perceiving.

That's why an effective affirmation is stated in the present tense. Let's say you have scheduled a meeting with a client to whom you are offering your consulting services. A good affirmation would be, "I am so excited about this project! It is the perfect opportunity to express my creativity and purpose for my business. I am successful."

Avoid the use of any negative terminology. For example, "I don't worry about money; I won't have trouble paying my bills; I can't fail at my business." This wording has completely the wrong focus. The action words are don't, won't, and can't. The subconscious emotional triggers are:

- **Worry about money**
- **Have trouble paying my bills**
- **Fail at my business**

What does saying those phrases bring to mind for you? They probably make you visualize some pretty negative scenarios. Remember, it's important to keep your mind *off* things you don't want.

Get it Right

You can use **positive affirmations** centered **in the now** to start your morning and to trigger uplifting emotional states throughout the day. Here's one for creating a wealthy mindset.

"I am experiencing ever-increasing financial prosperity. Money is supporting me continually and abundantly in fulfilling my Life's Purpose."

In Your Mind's Eye

Affirmations that you say out loud (and it is suggested you *do* say them out loud) are an outward expression of your creative desires. Visualization is an inner expression of your creative desires and is absolutely essential for the creative process. It is how you learn to feel your goals deep down inside, as if their results are already part of your life. These two tools (affirmation and visualization) are what you use to tell yourself a new story about your life.

Perhaps you think this is simply an attempt to wish away reality. It isn't. Did you know that your memory is so selective that how you currently view your past (and present) is already just a 'story' that you tell yourself? Compare your memories with someone else who was present at key points in your life and see how much your recollections differ.

You will quickly discover that your current memory has no more relation to 'reality' than any other perspective. So, making a new story isn't cheating, it's taking charge of the power and responsibility you have to create your life as you want it.

Write it Down

One really awesome way to create a new story is to actually write it down (just as we did at the beginning of this book). Flesh out your desire in as much detail as possible and put together a story of what everything looks like when you have achieved it. (Remember to state it in the **now**.) Cover all the ways it has changed your life; your business, relationships, finances, health, and personal growth. Besides writing the story, you can also record it on audio and play it at night. This way your subconscious can continue to absorb it as you sleep.

Be Committed

Once you create your new story set aside at least 45 minutes a day to spend on visualization (this can be divided into three 15 minute sessions). Are you telling yourself a negative 'story' about how you don't have time to do this? That's just an excuse; you have plenty of time to fill your life with a reality you aren't happy with! It's time to make a different choice. You will be glad you did.

Use your visualization session to build a clear picture of your new, positive story in your mind. Put yourself in the role of the main character, not a spectator. Become fully immersed in the imagery and it will be easier to stay focused.

Imagine with all your senses:

- Hear the sound of applause as you present your business ideas to an enthusiastic crowd.
- Taste the cuisine at a 5-star restaurant as you treat yourself and all your friends to a delicious dinner.

- Smell the scent of pine trees on the 20-acre plot of land where you are building your dream home.
- See the giant sum you are writing on a check for your favorite charity.
- Touch the fabric of the designer dress you are wearing to an important social or business event.

Exposing yourself to aspects of a wealthy lifestyle, even in small doses, can come in really handy during visualization. The more sensations and experiences you have to draw on, the more vivid and realistic these internal scenarios will be (they will also be more fun!). So, be sure to treat yourself to the better things in life from time to time and thoroughly savor these experiences.

Clarifying Wants vs. Desires

Before you read Chapter 10 about goal setting, it is important for you to stop and think about what you truly desire. Remember that **action** is a critical part of the **attraction** process. You will not work a lifetime to satisfy your wants; there is just not enough motivation there. This is why you see people in high-powered jobs getting burnt out and having health problems. They may have the money to pay for all their 'wants,' but they are sacrificing what is really important to them, pursuing their dreams, spending time with family, etc.

You **will** have the motivation to pour your life into your values. These are the things that express what you truly desire. When you set goals they need to be in alignment with your values and your passion. Once you have dealt with your internal money issues, there need be no conflict between attracting prosperity and living your values. You will simply be motivated to create goals that involve true abundance (health, wealth, and success) rather than just money 'at any price.'

Alisse's Example: I have never been madly motivated by money itself ... but I am very motivated by living a life with MEANING. I absolutely love that I live a life surrounded by beautiful things and I can experience the world in so many exciting ways.

However, I deeply love that I finally have the freedom to parent my children. I made a strong decision very early in my life (even before I had them) that I would be there for them always. Now I have that. I have Sunday off, work from 9am-3pm (school hours) during the week, and sometimes again later at night (instead of watching TV). I love what I do for a living, so this is easy.

This was always one of my highest priorities. So I made a ***decision*** to get good at creating and attracting money and then finding ways to diversify my skills and create streams of passive income. It has been the trial and error that I have experienced in my own personal growth that has led me to search for and develop the concepts that are the secrets of my success. Focusing on what I *truly* value has fueled me to do this.

Further to this, I have clearly identified what my values are in my business. For example, I have been able to ascertain that three of my greatest business values are connection, authenticity, and freedom. Whenever I go into any business / financial venture or any partnership I take these values in with me. If I feel it aligns then I progress; if not, I realize that no matter how good the opportunity might seem on paper, I am not going to thrive in this environment, so I gracefully decline.

Moving to the Next Level

As you start creating a new story of health, wealth, and success in business, you may notice areas of difficulty or obstruction. That's a normal part of the process. Each time you come up against a barrier, review the following checklist to see if there are changes you can make in one of these areas. Each item on the list can apply to your business, personal life, health, relationships, etc. These are phrased as questions so you can practice using your intuition to find the answers. Along with each question, you will find some Feng Shui tips and helpful colors.

BREAKTHROUGH TO SUCCESS CHECKLIST

1. Have I taken complete responsibility for where I am right now?

As long as you feel like the victim of circumstances, you will not claim your authority and ability to create success. Be on the lookout for thoughts such as, "There's no way I can increase my business when the economy is this bad." Instead, be thinking: "What opportunities does the current economy offer for me to increase my business success?" A further Feng Shui exercise for your language is placing the words "I choose" in front of your daily tasks rather than "I should ..." or "I have to ...." An example of this is, "I choose to pay my business taxes." Or, "I choose to work two extra hours tonight so that I can spend all day with my family tomorrow." Reaffirming to yourself that everything you do is your choice connects you with the reason you are doing it and directs your focus to your goal.

Color: navy or dark blue – taking personal responsibility.

2. Am I 100% clear about what I desire?

If you aren't used to thinking about what you really want (without limitations), this step can take some time, though it is time worth investing. In fact, you need to be able to tell the whole story of how you want your business and life to be, as accurately and vividly as you can state your own street address, making sure that it madly excites and inspires you in every way. Write it down in a special journal, (read it onto an audio that you can listen to), make a Goal Board with pictures, and place symbols that anchor in your goals throughout your entire office. These will assist you in visualizing your goals repeatedly throughout the day.

Color: yellow and gold – inspired thinking.

3. Am I really ready, with my systems well established?

Systems are a vital part of good Feng Shui in business. You may have concluded that Universal Laws teach us to put in place our goals and then sit and wait for them to turn up. While this is a very significant part of these teachings, they do not rule out the need to build solid systems as well. As you have learned, Feng Shui works very much in line with the earth's elements. Be aware that everything in the Universe is only as solid and optimal as its foundations and structure. In business this includes the systems you have in place and the plans you make toward achieving a goal. These save you time, simplify the operations of business, give you the means to make accurate changes where needed, and give you a clear picture of where you are at in achieving each goal. For example, you cannot expect business success if you have no idea how much money is coming in or going out. Have a solid financial recordkeeping system so you know how much money you need to reach your goals.

Color: white or metallics – structuring systems.

4. Am I staying completely focused on my plan?

It's easy to be distracted by irrelevant stuff that doesn't really bring us closer to our goals. That's another reason to clarify your goals / plans so you don't spend your time chasing 'shiny objects' that are not in line with your specific vision. In the same way, you mustn't let yourself be distracted by negativity. Avoid talking about negative topics with your friends. Instead, talk about how you desire things to be and your excitement over the future. Talk about it as though you have already achieved it. Again, turn off the radio if it is filling your environment with bad news. Also, don't get sidetracked by trying to get everything perfect. Gratitude and success are strongly linked, especially the attitude of gratitude in advance. This means you acknowledge your infinite potential and are already thankful for knowing that your goal is on its way. To help you focus, you can make a gratitude journal where you list what you are thankful for. Add your current goals as well and on the opposite page record three tasks you will complete that day to bring your desire into your reality. (We will look at this in more detail at the end of this book.) This is good Feng Shui, since it helps you set intentions.

Color: yellow and red – staying focused.

5. Do I really believe I can?

One of the key ingredients to achieving a goal is the belief that you can. This may be the most challenging part when a lack of belief seems impossible to overcome. Often this is because we think belief is something that you either have or you don't. It is important to understand what belief really is. Lack of belief or 'faith' is not an elusive phenomenon over which we have no influence. We simply have a tendency to believe what we are already familiar with. When you have difficulty believing a new concept, this is often a matter of simply not having the right neurological pathways laid down yet.

Repetition seems to be the key to solving this. Repetition of an affirmation over and over to yourself (as long as it is not counteracted by the repetition of an opposing affirmation) will build new neurological pathways and alter your belief in yourself.

The second key is direct experience. Once you experience something new the body very quickly lays down neural pathways which support that behavior. For example, you may have a goal to transfer all your business procedures to the computer. Your initial thought might be, "I can't use a computer well enough." This only represents your current neurology (any fear you feel is simply that as well). You can change this by affirming your desired belief and continually repeating it to yourself. Do this verbally and through all the other ways you have learnt to use your environment with Feng Shui.

The next step would be to attend an introductory computer class and learn the first step. The first time is awkward and uncomfortable because you do not have matching pathways in your brain. However, your mind will very quickly figure out how to do this. Then, it is a building process. With every experience a new belief is accepted. You have a mind / body / soul that is eager to change its knowing in any moment.

Here's a great Feng Shui tip for changing your beliefs. Save an empty tin can. Take a piece of colored paper or felt and cut out a shape of an eye. Stick the eye to the can and place it where you will see it often. Fill it with soil and plant a seed in it, then nurture its growth daily. This is your very own 'I Can' can. Whenever you see it, you will be reminded that you can do anything you set your mind to.

Color: purple – belief in self.

6. Am I neglecting what I truly value in life?

This concept is closely linked to your goals, but it is more about your values. We have already broached the subject of values, though now it is your turn to think about this in relation to your own situation. Are you sacrificing what you really want deep down inside to reach the goals you think you want? For example, you might be pursuing a goal for the purpose of making a larger passive income stream. Your underlying purpose might be the freedom to attend your children's school events. Yet you are so busy and overwhelmed in chasing your goal that you hardly see your children at all. Sometimes, we sell ourselves the notion that we will give up what we want now in order to benefit later, and yet this never seems to work. For one thing, the subconscious mind does not keep time and only relates to what it is surrounded with right now. In addition, you may have the determination to keep up the fight for the goal, and yet your motivation will fade as you get burnt out.

This is why we see the scenario so often where an illness or injury stops someone in their tracks when they are pursuing some unhealthy vision that does not match what they truly desire deep down. This sort of goal hunting almost never works and even when it appears to be providing the results, actual fulfillment tends to evade these people.

The way to materialize your desires quickly and effectively is to think how to introduce the **way** you want to live and then add these features to your business life now. If you would love to have the time and money to travel the world, then go somewhere each week or so, even if it is just down the road. Or promote your business in areas you would like to visit and make a point of going there. Decide that you **do** travel and your energy and excitement will grow, attracting more of the same. Don't hold your breath believing that one day it will all just change. It actually won't. An interesting exercise is to ask yourself seriously, "What would I do if I only had 30 days to live?" Start to bring into your business and life what is really important to you right now. Be innovative.

Color: turquoise blue – true to yourself.

7. Am I giving enough service?

The amount of money you earn will always be equal to the amount of service you give. This is based on the Law of Cause and Effect. Ways to give more service include having a greater understanding of the needs of others and providing their solutions, committing to continually get better at whatever it is you do, and improve the positive impact that you have on others. Ask yourself this question: would I be missed if I were not here? How long would people miss me? This will give you a good idea of how valued your services are. Think of new and exciting ways to serve your clients better. This can be little things like giving beautiful gifts, sending positive words, offering them a tasty snack, showing them how to use a tool that will save them time / money, etc. The more you give from a place of abundance, the more you will receive.

Color: fuchsia – Universal love and service.

8. Have I asked for the help I need?

Sometimes, you really do need assistance to move forward and meet your goals. It is interesting that we attempt to achieve exceptional results in our business without employing exceptional skills. Big business never does this. It is also unrealistic to expect ourselves to have the diverse range of skills necessary to build a successful business from financial management, to bookkeeping, to website building. Be prepared to make an investment in your business by hiring an expert to give you the training you need to reach the next level. If you don't feel comfortable investing money in your business by paying someone else for their expertise, go back through the previous questions on this list and figure out why.

Locate the northwest sector of your office. Place a symbol of a turtle there as this invites helpful people to your business. This helps you find the exact person you are looking for.

Color: yellow and turquoise – the ability to be able to ask for what you need.

9. Is there a greater lesson in all of this?

What more do you still need to learn in order for you to let go of this current 'challenge?' It is a very powerful transition that takes place when we embrace our 'life lessons,' and it is interesting how they keep turning up when we don't figure this out. Accept this as your journey and ask the Universe, Divinity, God, your Angels, or your Subconscious Mind to show you what you are supposed to be learning from events in your life. Write a note with your question on it and place it under your pillow.

Color: multiple colors (try opal) – diversity.

10. Is my old programming out of line with my goals?

Limiting beliefs and decisions are always one of the most common reasons for a lack of success. To overcome negative programming, sit in a quiet space with a quiet mind. Reframe the beliefs you have so that they support your goals. Use repetition of positive affirmations to accomplish this gently and effectively. Remain aware so that you can identify and replace old beliefs until you have de-cluttered your mind completely. A Feng Shui technician can assist you in uncovering hidden beliefs that are holding you back. Then, you can make changes in your outer environment that reinforce your positive affirmations and raise your vibration to match who you want to be.

Color: orange – letting go of old programming.

LIGHT A CANDLE TO SUPPORT YOUR GREATER INTENTIONS

Now that we have looked at how to place our intentions into our outer environment (anchoring) and how to turn them inwards to our 'internal mindscape' (affirmations and visualization), it is time to learn how to send them out into the ether for Universal support. A candle is the perfect Feng Shui tool to do this.

The hidden truth of our ability to materialize **anything** we want is that this is a collaborative effort of our own creative ability plus that of the Universal energy supply. This makes sense when we realize that as we raise our own vibration (improved thoughts and feelings) we are having to source this *extra potential* (inspiration) from somewhere.

Being aware of this *combined creative process* is the first step in having the ability to introduce more of Universal abundance into our business and lives; the next is to find a practice that encourages us to do this on a regular basis. That's why a candle is so ideal. It instills our commitment to embrace this advanced understanding and then do what is necessary to accept a greater energy frequency (more of what makes us feel good: wealth, joy, success) into our own vibration.

Once you have clearly articulated what you *intend* for your business, you can use a candle to assist you in overcoming challenges, finding answers, and harnessing greater help. Lighting a candle gives you a chance to 'make a wish.' The size, shape, and color of the candle needs to support the energy of the intention you desire to create.

Directions for Using Candles:

1. The color of the candle and its vibrational force is released into the atmosphere when the candle is lit.
2. Choose the candle color, shape, and size based on your *wish*.
3. Have a definite intention when you light the candle. Sit for a moment focusing on this before you light it.
4. Once you have used a candle for a specific purpose, do not use it for another purpose. The energy of the candle will become confused.
5. Extinguish the candle with a small metal cup or candlesnuffer. Avoid using your breath as this is a creative force and so is the process of working with the candle.

Note: Always be safety conscious when using a candle. Keep it clear from flammable objects such as paper and curtains and be sure to put it out once you have completed this process. It is not advisable to leave an unattended open flame.

What Shape Shall I Choose?

Egg Candles: Use an egg-shaped candle when you desire to create something new in your life.

Round / Cylinder Candles: Use round candles when you desire to complete something.

Square / Rectangle Prism Candles: Use square candles when you desire to bring security to a situation.

Extra tip: Use different colored candles and candle-holders or, alternately, place different colored pieces of paper under your candles to benefit from the power of the color combination. Refer to Chapter 9 for the personal color combination table.

What Size Shall I Choose?

The size of the candle matches the amount of time you desire to spend with the wish. (How long you think or want it to take to materialize.)

What Color Shall I Choose?

Red: Use red when your *intention* is to create new business goals and projects or inject energy into existing ones.

Orange: Use orange when you are unsure of which direction to take or your *intention* is to make big changes in your business.

Yellow: Use yellow to stimulate great ideas, have more clarity, and make good decisions in your business; also use when your *intention* is to raise the positivity and happiness in your office space.

Green: Use green when you feel stagnant and your *intention* is to create business growth and to improve your foundations and systems. Green enhances balance.

Turquoise: Use turquoise when your *intention* is to set better boundaries in your business and honor what you truly desire.

Purple: Use purple for any business venture where your *intention* is extra strength and belief in yourself.

Pink / Fuchsia: Use pink / fuchsia when your *intention* is to enhance your connection with what inspires you about your business and to share this passion with the community.

White: Use a white candle when your *intention* is to take some time out or purify your space.

Candles, Affirmations, and Visualization

A candle is also the ideal companion to use with your affirmations and visualization. This is because using a candle:

- Sets your focus
- Calms the mind from anxious or negative thinking
- Gives you a start and finish time from being lit to extinguished
- Raises your vibration to one of joy and optimism
- Is regularly associated with events in your life where you have been made to feel valued and significant
- Anchors in the successful outcome of your desire

Client Story: Zoe is a client who started her own business as a sales consultant. While she was very excited for her new opportunity, she was also very concerned that her motor vehicle was not going to be adequate for the job. Zoe needed to drive a lot to be successful in her new career and her old car had mechanical issues and was unreliable. She also felt that its condition and age was not the best image for a successful business owner, and wondered if her clients would question her credibility when she arrived in it. While Zoe needed a new vehicle she did not have the financial capacity to access one at the time.

I suggested that Zoe write her desire as a powerful affirmation stating that she already had a new vehicle. She wrote, "I am so thrilled and thankful now that I have the exact car I need for my new business. I love it and feel strong and confident when I arrive to meet with my new clients. I feel like a successful businesswoman and I am now eager to share my product." I then suggested she purchase a red egg-shaped candle, write her affirmation on a green piece of paper, and find a picture in a magazine of a motor vehicle she loved.

Each day I asked Zoe to read her affirmation and look at the image she had chosen of her new car and then place them both under the candle she had bought. I had her light the candle and spend 15 minutes three times every day visualizing herself driving her new car. When Zoe visualized I suggested she use all her senses: see herself walking to and sitting in her new car, hear the sound of the engine as she started it, smell the fresh new interior, feel the steering wheel in her hands and the soft seat covers underneath her. I also suggested that Zoe go down to a car yard and test drive a brand new car, just so she could heighten her senses when she visualized the process. I asked Zoe not to spend any time trying to figure out how she would **get** the car, as this would only detract from the exercise.

What happened next was very interesting, though not uncommon of the process. Within three weeks of doing this exercise on a daily basis Zoe received an unexpected phone call from her dad. Zoe had not grown up with her dad and had only seen him a handful of times in her life. Due to her dad's health he had decided to move into a retirement home. He asked Zoe if she wanted some of his possessions he would not need at the home, one of these being – you guessed it – his car.

He had bought it only three years earlier, maintained it well, and it had hardly any mileage on it. It was the exact color of the image of the car Zoe had placed under her candle. He didn't want payment for it; he simply wanted Zoe to have it. All Zoe needed to do now was accept the gift and she had the car she had asked for. She also took the opportunity to build a relationship with her dad and now spends time with him each week. Zoe is very grateful for the events that came to fruition and she is going really well in her new sales position.

Other Client Examples: Each of these clients followed the exact same steps relevant to their own business desires. Here are their outcomes:

Debbie decided she wanted to refurbish her nail salon and received an unexpected check from the tax office. **Bernie** decided he wanted to double the sales of the office furniture he produced and was contracted by a local law firm to do a complete fit out for their new building. **Kym** decided she wanted to begin facilitating workshops for parents on *good nutrition for lunchboxes* and secured a joint venture with the local health food store. She filled her first event. **Michelle** dearly wanted to take a holiday which she couldn't afford until one of her clients was so thrilled with the art work she supplied they unexpectedly offered her a weekend at their beach house in gratitude for doing such an amazing job.

The materialization process really is incredible. It works like magic. I have witnessed it time and time again with my clients as well as in my own business / life. See how many ways you can use it to create and attract (manifest) the change you want.

FENG SHUI TO OVERCOME BUSINESS CHALLENGES

Sometimes beliefs and internal attitudes aren't the only issues creating bottlenecks in your success. Behaviors and habits are other factors that play a significant role. You keep doing things a particular way out of habit and this impacts your business in ways that you don't even realize until you take a closer look. There are some common business challenges that turn up for many entrepreneurs, time and time again, and when we approach them the same way we continue to get the same result. What if we approached them differently?

Here are Feng Shui remedies for some common business challenges. These integrate what you have learned so far and give you another way to overcome issues in your business. You've seen many (but not all) of these tips in previous chapters. Here, they are arranged specifically by the challenge they are designed to address. They each include a Candle Ritual to assist the process. This is a great list to keep on hand for when you are feeling stuck or overwhelmed by a persistent challenge (or you just want to create something miraculous and new).

Step One: Take a moment to sit in your workspace and experience your surroundings. Then ask yourself if you 'feel good' in this space.

Step Two: Identify anything in your environment which reflects and confirms the current challenge.

Step Three: Create your environment to reflect your new goal.

Note: The Candle Rituals included with each business challenge integrate an affirmation, visualization, and specific shape / colored candle to work with. They also direct you to the best space in your office to implement them. Keep in mind if it is not suitable or appropriate for you to always use a candle you can still do the affirmation / visualization component of each of these procedures without one. Waiting at the airport, sitting on a park bench, or taking a few moments in a parking lot are all adequate spaces to nab your time to do this. In fact, once you have completed these procedures a few times **with** a candle, your mind has the ability to recall and recreate this same vibration no matter where you are. That is why working with energy-based strategies is so beneficial; they just keep expanding.

Now let's bring it all together ...

FENG SHUI FOR COMMON CHALLENGES:

1. Not making enough money ...

Decide on an affirmation that you will know by memory and repeat this regularly. Here is the earlier example you can use:

"I am experiencing ever-increasing financial prosperity. Money is supporting me continually and abundantly in fulfilling my Life's Purpose."

Now sit in your office space and look around you. Ask yourself these questions:

- Does your environment look beautiful and abundant? Does it look like the office you would have if you were making the money you desire to make?
- Do you have any old or broken items in your office? Does all of your equipment work efficiently and effectively?

- Are your bills visible from your desk e.g., hanging up on a notice board / in a slot that you see numerous times per day? Are receipts lying around in plain sight?
- Do you have pictures around you of rough seas, desert scenes, or empty boats?
- Do you have any stagnant water in or outside of your office such as a dirty fish pond or a scummy rain barrel?
- Do you have empty containers in your office?

Implement these Feng Shui strategies to address this challenge:

- Be sure your environment looks abundant and wealthy. Add features, symbols, and colors that represent wealth. Make sure there is a good balance of the water element present.
- Clean and de-clutter your office. Remove any trinkets or other items that do not mean anything to you and do not feel abundant. Repair or throw out any broken items.
- Check the 'Red Phoenix' of your office; remember that you will only invite as much abundance to you as you demonstrate at the entry of your business. Add a frog symbol to the front entry of your office. This can be a Chinese Money Frog or any frog symbol that complements your specific interior design. The frog is a luck symbol for bringing money in. Display it at hip height or above and be sure it is facing inwards. (A tortoise symbol at the back door will stop money leaving your business too quickly.)
- Get your filing organized. Put your bills away and file your receipts so that they are ready for you when you need to do your books.

- Be sure you are sitting in the power position in your office (reminder: in the center of the room, facing the door).
- Replace pictures which represent lack, difficulties, and abandonment with abundant pictures. Clean and freshen any water around the office. Check the positioning of all fish tanks, water fountains, and water symbols that they are not hindering your finances. Remember, water represents how and where your money is flowing.
- Place something in any empty pots. Try decorative bamboo, silk flowers, or even gold nuggets (paint pebbles gold to represent these).
- Activate the **southeast** (Business Wealth) sector of your office using the Bagua instructions and enhance your **Success Direction** with the suggestions provided in the preceding chapter.

Bonus Suggestions:

- Place a Chinese Red Envelope (this is a small red envelope with Chinese luck symbols printed on it) in your wallet with 3 dollar coins (or bills) inside it. This will attract money to you. You can also place one of these in your cash box or diary if you wish. Place a small mirror in your cash box / register to 'double' your money. *Follow the '10 hot tips to Feng Shui your wallet' you will learn in Chapter 9.*
- Place a set of 3, 6 or 9 Chinese Coins on the inside handle of your office door. Face the yang side of the coin (with the 4 symbols on it) inwards so that you can see this side. Tie the coins with a red ribbon with gold trim; secure them so that they aren't swinging as you open and shut the door (a store familiar with Feng Shui will stock these).

- Put your bills into a green folder with a Chinese Red Envelope inside. Also, place a red or green piece of paper inside the folder with your financial affirmation written on it. File the folder in a set place in your cabinet that is below hip height to reduce the number of bills you are receiving. (Storage at hip height represents the present, below – the past, and above – your future.)
- Place an empowering picture on your bills folder. This needs to be one which immediately makes you feel like you have the power to do anything. This may be a picture of you when you achieved something and felt very powerful, an image of a mentor who makes you feel this way, or anything that looks like increasing money or wealth. This is a great way to raise your vibration so that you feel empowered to handle money matters.

Candle Ritual for Attracting More Money:

Purchase a **green rectangular-shaped candle.** Write your Attracting More Money Affirmation on a red sheet of paper. Take a picture of money, actual notes of money, 3 X $1 coins, or a print of a check with an amount of money you desire on it (sign it from the Universe) and place it under the candle with your red sheet of paper.

Each day light the candle and say your affirmation out loud. Then spend 45 minutes (or 3 x 15 minute sessions) visualizing what your life looks, sounds, feels, smells, and tastes like having attracted your desired amount of money. After this time put the candle flame out. Do this each day until the candle is finished. Journal your experiences and the changes in your *money* state that occur. Perform this exercise in the **southeast** or **Success** sector of your office.

2. Not attracting enough clients ...

Decide on an affirmation:

"I am easily able to attract an abundance of clients to my business. Clients seek out my products and services and love the experience I offer."

Ask yourself these questions:

- Is your front door able to open fully? Is it easy to open and used regularly? Do you have ample space in your waiting room to fit more clients?
- What is the energy of your Business Branding? Is it dynamic and eye-catching while being solid and professional? Is it consistent? Have you used appropriate colors for the market you are attracting? What is the impact of your website homepage?
- What is the signage on the front of your office like? Is it visible from the road and in good proportion to your shop? Is your front entry (gate and door) to your office / workplace pleasant and inviting? Is it beautiful? Is your office / workplace itself easy to find and visible from the street?
- Do you have a diary, and is it big enough to fit all of the appointments you desire to have?
- Is the ring tone of your business phone pleasant? Are there things in your office which represent aloneness such as single chairs, single items, or pictures of solitary people?

Implement these Feng Shui strategies to address this challenge:

- Check your branding. It needs a clear message, an eye catching design, appropriate colors for your market, consistency, and professionalism (everything you have learned in chapters 4 and 5). If you are sure of what you offer, your clients will be also.
- Invest in your website design. This is a 'virtual' front door to your office. Use consistent branding here as well as other tips like colors, layout, pictures, and the right size header for the site. Don't use pop-ups for offers and selling products. These are an obstruction to enter and flow through your website. If you intend to include these have them appear only when the viewer decides to leave the site.
- Your front entry is vitally important again. Any time you desire to *invite more of anything* in your business, check your 'Red Phoenix'. Make sure it is visible, inviting, abundant feeling, clean, in working order, easy to enter, and not obstructed. Check that the hinges on doors and gates are oiled. (For home-based business this is the house's front door / gate as well as your office door.) Wipe them down to refresh the energy. The door needs to be able to open **fully** (don't hang or store anything behind it). Place healthy plants at the front door to invite clients in.
- Also for office buildings your front sign needs to represent your clear branding, be pleasant and proportional to the size of your business, and able to be clearly seen from the street.
- Be sure your office space is inviting even if you do not see clients face-to-face. Use colors such as yellow, orange, and pink to invite people in.

- Remove anything from your office which represents aloneness such as single chairs, single items, or pictures of solitary people. Add pictures to your office space, screensaver, and mobile phone of two or more people or animals to represent good communication and business relationships. The symbol of a ship with 3 x gold coins on it placed in the top left-hand corner of your desk will attract wealthy customers.
- Be sure you have enough space in your office and waiting room to accommodate an increase in clients. For example, you need plenty of chairs in your lobby as well as space in your filing cabinet to expand.
- Have a diary that is large so that you have plenty of space to write appointments for extra clients and expanded business activities. Choose a vibrantly colored diary. A red diary with an orange piece of paper used as a book mark is powerful and ensures you always have enough clients. You may even wish to write your affirmation on the orange piece of paper:
- Have a ring tone which makes you feel great when you pick up the phone. Your energy will be transferred to the client. Be sure when you make phone calls to potential clients that your energy is high and you are feeling empowered.
- Activate the **southwest** (Business Relationship) sector of your office from the Bagua Map provided.

Candle Ritual for Attracting More Clients:

Purchase an **orange egg-shaped candle**. Write your Attracting More Clients Affirmation on a fuchsia pink or red sheet of paper. Place a picture of people interacting, a pair of anything, birds communicating, a world map or anything that represents the earth (including crystals), under the candle with your fuchsia pink or red sheet of paper.

Each day light the candle and say your affirmation out loud. Then spend 45 minutes (or 3 x 15 minute sessions) visualizing what your life looks, sounds, feels, smells, and tastes like having attracted your desired number of clients. After this time put the candle flame out. Do this each day until the candle is finished. Journal your experiences and the changes in your *attracting clients* state that occur. Perform this exercise in the **southwest** sector of your office.

3. Not keeping clients ...

Decide on an affirmation:

"Clients come easily and readily to me. They love what I offer and are drawn to come back time and time again."

Ask yourself these questions:

- What is the first thing your new client sees when they walk in through your front door? Does it feel uncomfortable to sit in your waiting room or office?
- Do you feel productive and motivated in your office, or tired and sluggish after working there for a period of time? Do you have a clock directly in front of you when you walk into your office or sit at your desk? Is your office space too yin and needs more yang energy?
- Do you have a lot of single items in your environment? Do you have a lot of pictures in your environment which represent aloneness or lack?
- Is your business filing system and paperwork management system easy to work with, clear, concise, and up to date?

- When someone enters your online office (website), is it clear and consistent in its opening message, does it direct them to their next step, and is it easy to navigate?
- Are the emails you send out pleasant to look at, clear to read, and direct in their message and next action step?

Implement these Feng Shui strategies to address this challenge:

- Check the first thing your new client sees when they walk in through your front door. Be sure their experience when they enter your environment matches how you want them to feel. Make sure your waiting room is inviting and comfortable to sit in. Attend to sounds, smells, temperatures, and treats. (Don't have outdated magazines and irrelevant reading materials on waiting tables. Be deliberate here also.)
- Create dynamic flow in your office space and online office (website). The client needs a consistent message in your office that matches the one they learned from your branding. Everything your client connects with must feel clear in its message. Your website needs to be beautiful and allow the client to flow easily from one step to the next. Be sure that images load quickly.
- Any emails you send out or correspondence you post must be pleasant to look at, clear to read, and direct in the message and next action step. Make sure when you communicate with your clients via any technology the reception is good and clear (mobile phone, tele-seminar, webinar). If you are using face-to-face technology (Skype, Zoom, Facetime, audio, video, etc.) be professionally dressed and ensure good picture and sound. Give clear instruction to your clients as to how they can successfully use these facilities.

- Be aware that flow in your office space can be blocked by clutter and mess, broken and worn items, old and tired looking furnishings, and neglected plant life. (A client will feel this whether they are actually visiting your office or just dealing with you over the phone!) Blocked flow is definitely transferred in everything we do. Check your workspace colors and décor for the yin-yang balance. If it is too yin, then clients will not be attracted to do business with you. Alternately, excessive yang will mean they come and go too quickly.
- Have items in pairs that are positioned as though they are having an enjoyable conversation. Have pictures with two or more people or animals interacting pleasantly. Make sure your images are joyous and positive looking.
- Be sure your business filing system and paperwork management system is clear, concise, and up to date. You will unconsciously repel clients if you feel bogged down at the back end. A green diary with a blue piece of paper encourages consistency in business growth.
- ***Poison Arrow*** *(negative Feng Shui)*: A clock on the wall opposite the entry, even in a home office, will mean clients do not stay long.
- Activate the **east** (Business Health) sector of your office from the Bagua Map provided.

Candle Ritual for Keeping Clients:

Purchase a **green rectangular-shaped candle**. Write your Keeping Clients Affirmation on a blue sheet of paper. Place a picture of a forest, a thriving tree, a bowl of fruit, a bunch of flowers, or a symbol of a monkey under the candle with your blue sheet of paper.

Each day light the candle and say your affirmation out loud. Then spend 45 minutes (or 3 x 15 minute sessions) visualizing what your life looks, sounds, feels, smells, and tastes like having mastered keeping the clients you want. After this time put the candle flame out. Do this each day until the candle is finished. Journal your experiences and the changes in your *keeping clients* state that occur. Perform this exercise in the **east** sector of your office.

4. Lack of clarity in your direction ...

Decide on an affirmation:

"I am clear and purposeful in my business direction. I know exactly where I am going and what I desire to achieve."

Ask yourself these questions:

- Do you have adequate light in your office space? Are you able to see out from your office (say, through a window)? Are your windows clean?
- Is your computer screen clean? Are the keys to your office, house, or car tidy? Is your car clean, serviced, cared for, and beautiful to you?
- Do you have a Goal Board with your dreams and desires on it in full view from your office desk? Have you set clear concise goals for your business both long term and short term?

Implement these Feng Shui strategies to address this challenge:

- Improve the light in your office via natural light, turning on a light, adding a lamp, or even lighting a candle. Avoid the use of glaring or unpleasant lighting.

- Clean your office windows and your car windows as well. Clean your computer screens, mobile phone screens, eyeglasses if you wear them, etc.
- Your car represents your direction and where you are going. Be sure it is clean, serviced, de-cluttered, well cared for, and appealing. Even if this is not your ideal car, you can still make it beautiful to you. Our gratitude for what we have opens us up to receive more.
- Feng Shui the keys for your office / house and car by adding a lovely key ring holder. Have all your keys face the same direction. Remove old, broken, or unused keys. Have keys all connect by one ring as opposed to ring connected to ring. Remove any unnecessary links.
- Clean your shoes and organize them in your wardrobe with the toes facing out (as described in the next chapter).
- Write your goal for clarity on a yellow piece of paper, put it in your Success corner and then place a compass on top. The compass will help you find your direction. If you think you still need more information in order to make the best decision use the symbol of an owl.
- Spend some time deciding what your goals really are (long and short term) and **write them down**. Also, create a Goal Board which represents them visually. Be sure you can see this board when sitting at your desk.
- Activate the **north** (Career) and **south** (Business Fame) sectors of your office from the Bagua Map provided.

Candle Ritual for Clearer Direction:

Purchase a **yellow cylinder-shaped candle**. Write your Clear Direction Affirmation on a red sheet of paper. Place a picture of a sunrise, birds flying, a compass, an article of recognition, a phoenix, an owl or anything that represents your ultimate business goal, under the candle with your red sheet of paper.

Each day light the candle and say your affirmation out loud. Then spend 45 minutes (or 3 x 15 minute sessions) visualizing what your life looks, sounds, feels, smells, and tastes like when you have mastered your absolute clear direction. After this time put the candle flame out. Do this each day until the candle is finished. Journal your experiences and the changes in your *clear direction* state that occur. Perform this exercise in the **south** sector of your office.

5. Not enough time ...

Decide on an affirmation:

"Time is abundant for me. I always have enough time to fulfill all that I desire to be, do, and have."

Ask yourself these questions:

- Do you have an abundance of clocks and calendars in your workspace? Is there a calendar directly in front of you where you sit or as you walk into your office? What is the picture connected to your calendar and how do you feel when you look at it?
- Do you regularly use promotional material from other companies in your work and personal environment; e.g., pens, notepads, cups, t-shirts, etc.?

- Do you have more than one project on your desk at the same time even when you are not currently working on them all? Is your office crammed in or cluttered?

Implement these Feng Shui strategies to address this challenge:

- Avoid having too many clocks and calendars in your workspace especially right in front of you where you sit or as you walk into your office. (Check calendars on fridge magnets, computer screens, advertising material, etc. You may have more around you than you think.) Check what the picture connected to your calendar is and how this makes you feel when you look at it. Storm scenes and drought scenes (even promotional calendars) do not relate well to your feeling of lack of time. Instead choose a calendar with a picture that makes you feel calm (like flowing water) or empowered (like a sunrise).
- Place your calendar somewhere you can easily access it when you need to, such as in a handy drawer along with your diary (not in the bottom drawer). If time is an issue and you have it open in front of you constantly, this will make you feel pressured. Also take your watch off after you shower and at bedtime.
- If you do hang your calendar where you can see it, be sure it is above hip height so that there is more time in your future. Clocks and calendars below your hips make you feel like having time is a thing of the past.
- Be aware of using promotional products from companies (your company or any other company) all of the time. This makes you feel like you are always at work. Certainly do not sleep in t-shirts with company logos printed on them or use a promotional mug for your cup of tea before bed.

- Avoid pictures of holiday beach destinations where there is no one in the beach chair. This represents hope for a holiday you are never likely to take. Instead, have a holiday scene with people in it, or even Photoshop yourself into it.
- Only have on your desk what you are working on right now. File all other projects away and out of sight until you intend to work on them again. Store them behind closed doors. It only takes a few moments to do this and allows you to remain focused on the job at hand. Set specific times to answer emails and return phone calls, and abide by this schedule.
- Sit on a white or silver piece of paper / cloth when you are feeling really pressed for time. White gives you space to be productive under pressure. Silver helps you to be more precise.
- Activate the **west** (Business Creativity) and **northeast** (Business Belief) sectors of your office from the Bagua Map provided.

Candle Ritual for Having Enough Time:

Purchase a **white or silver cylinder or round-shaped candle**. Write your Enough Time Affirmation on a turquoise sheet of paper. Place a picture of people having fun, children playing, a light bulb, a clock, or something you have drawn or created, under the candle with your turquoise sheet of paper.

Each day light the candle and say your affirmation out loud. Then spend 45 minutes (or 3 x 15 minute sessions) visualizing what your life looks, sounds, feels, smells, and tastes like having materialized more time. After this time put the candle flame out. Do this each day until the candle is finished. Journal your experiences and the changes in your *more time* state that occur. Perform this exercise in the **west** sector of your office.

Note: Contrary to what you might feel, investing this time will give you more time.

6. No balance between personal life and work life ...

Decide on an affirmation:

"I have an abundance of time for work and play. I have a balanced schedule and easily stick to what my heart desires."

Ask yourself these questions:

- Do you have a set space in your home where you work whether you have a home-based business or you bring work home from the office? Do you leave paperwork or bills around the home such as in the living room, kitchen, or bedroom?
- Do you shut down your computer and business equipment at the end of the work day? Do you have a separate phone for work and pleasure? Have you allocated time to put your work phone away?

Implement these Feng Shui strategies to address this challenge:

- Balance is a decision that requires strong internal boundaries. You can see these reflected in your environment. Home-based businesses need to be very strong about this point. There needs to be a set office space where you can close the door once the day is done. Even if you work at another location and bring work home, this rule applies.
- If you cannot set aside a whole room for an office, use a work cupboard or cabinet where all of your work stuff goes and you can close the door. Physical boundaries represent emotional boundaries. Order and systems are vital for balance. Avoid

placing any invoices, quotes, contracts, work diaries or work phones in any space in the house other than the designated area.

- Avoid bringing promotional products with company logos into your home. If you drink from free cups, write with company pens, or wear promotional ball caps outside of business, you are carrying work around with you all of the time. You have crossed the boundary into your personal life. This is where you need a keen eye to check your home for any business intrusion which needs to go in the designated office space. Even your own company's promotional products must stay in the office space unless you are doing business or intentionally promoting at an event.
- Have a separate phone for your business or somewhere you can put your phone away after a set time. Even if you are not directly dealing with emails and work calls noticing them come in still plays on your mind.
- If you have trouble with shutting up shop, place a picture of your family / or your favorite hobby on your office desk in the right-hand corner; maybe a photo of the family having fun together, one that makes you feel good. Mark out the times in your diary when you will have family time, self-time, and relaxation time and stick to it.
- Activate the **center** (Chi Center), **east** (Business Health), and **west** (Business Creativity) sectors of your office from the Bagua Map provided.

Candle Ritual for Gaining Balance:

Purchase a **yellow square-shaped candle**. Write your Gaining Balance Affirmation on a green sheet of paper. Place a picture of your business vision, your 'how you want your business and life to be' story, a set of scales, the earth, a mountain, or a tortoise under the candle with your green sheet of paper. In this instance also add a piece of pink paper to balance the color sequence and ensure your desire is delivered in a gentle way.

Each day light the candle and say your affirmation out loud. Then spend 45 minutes (or 3 x 15 minute sessions) visualizing what your life looks, sounds, feels, smells, and tastes like having attracted the balance you desire. After this time put the candle flame out. Do this each day until the candle is finished. Journal your experiences and the changes in your *balance state* that occur. Perform this exercise in the **east** or **west** sector of your office.

7. Not moving forward, no business growth ...

Decide on an affirmation:

"My business growth is abundant. I am moving easily in my desired direction attracting all the results I have decided to receive."

Ask yourself these questions:

- Does your environment look tired? Is it in need of a fresh coat of paint and some new window dressings? Does it feel lifeless? Have you made any significant changes to your workspace in the past few years or has the environment become stale?
- Have you introduced beautiful color to your workspace, or is it mostly blacks, browns, and neutrals? Have you checked the yin-yang and 5 elements for balance?

- Do you have any stagnant water or dead plant life (including dried flowers) in or near your workspace? Do you have any images on display that represent the past or loss?
- Have you written down your goals and are they in a place where you can read them often?

Implement these Feng Shui strategies to address this challenge:

- Your environment may need a facelift. You must create a space that represents how you want your business **to be** rather than reflecting how your business was or is **now**.
- Your branding may need a re-creation. Freshen up your message with some new color and design layout. Print your business cards so that they are in portrait rather than landscape orientation (if suitable). This gives the feeling of upward growth. You can also create your Goal Board in this same way.
- Spend some time revisiting your goals (you will learn effective goal setting in the final chapter) getting clear on exactly what you want. Be sure you have written them down and that you read through them often. Make sure they are measurable, stating the exact amount of money and new and existing clients / customers you desire. Lack of growth is an indicator of too much focus on what you don't want and not what you do. Maybe it is even time for a new Goal Board.
- Have live plants in your office space. Nurture them and keep them abundant and growing. Variations of the color green, wood furnishings, and rectangular-shaped items all represent growth. Have a balanced amount of these in your environment. Check that all the 5 elements are represented in your work-space as we did in Chapter 6.

- Lighter and brighter colors are motivating while darker and more muted colors are more relaxing. Have ample color in your environment to keep the energy moving. The yin-yang balance information will help you here.
- Keep water fresh in ponds and fish tanks. Dust regularly and don't allow mold to grow anywhere. Add movement to your environment by opening curtains, hanging metal wind chimes, or selecting a moving screensaver.
- Images of tragic events, old and worn items, people who have passed or that focus on the past are not ideal for a thriving business (unless very appropriate for your business message). Black and white and sepia photos are for strategic use only. Excessive use of antique or second-hand furnishings can carry energy of the past. Items that are relevant to an old business or past endeavor need to also be carefully considered.
- One time per year move all furniture (even if it only 2 inches) from where it is sitting right now. As much as is practical move all of your furnishings and give them a good clean in and around. A good spring clean has a very positive effect on your business growth.
- Activate the **east** (Business Health) and **southeast** (Business Wealth) sector of your office from the Bagua Map provided. Activate your personal **Success Direction** as instructed in Chapter 7.

Bonus Suggestion:

- To attract more opportunities into your business fully clean out your work pantry and refrigerator. Do this for your home if home-based. Food represents nourishment and these areas store the resources you *create with* to achieve this. Place a picture of 3, 6, or 9 sunflowers on your refrigerator. Ensure they have green centers and not brown (indicating they are still full of life). Don't be surprised at how quickly **this** exercise works as new and exciting opportunities immediately fall into your lap.

Candle Ritual for Creating Business Growth:

Purchase a **green rectangular-shaped (tall) candle**. Write your Business Growth Affirmation on a turquoise sheet of paper. (Add to your note the exact amount of money, clients desired, etc.) Place a picture of a thriving tree or plant, a forest, a running stream, a giraffe, sunflowers, a frog or water creature under the candle with your turquoise sheet of paper.

Each day light the candle and say your affirmation out loud. Then spend 45 minutes (or 3 x 15 minute sessions) visualizing what your life looks, sounds, feels, smells, and tastes like having attracted your desired business growth. After this time put the candle flame out. Do this each day until the candle is finished. Journal your experiences and the changes in your *business growth* state that occur. Perform this exercise in the **east** or **Success** sector of your office.

8. Feeling overwhelmed and out of control ...

Decide on an affirmation:

"I am calm, clear, and centered. Every day in every way I allow my prosperous business to flow."

Ask yourself these questions:

- Is my office overwhelmed with clutter? Do I have structure and systems in place that make it easy for me to do business? Do I have adequate space and storage for all the items I need to do business? Can I access everything easily?

- Am I sitting in the power position with a strong supportive wall behind me? What is my office chair like?

Implement these Feng Shui strategies to address this challenge:

- De-clutter your environment. Include your handbag, your briefcase, trunk (boot) of your car, your desk top, and your office. Keep de-cluttering and organizing until clarity returns. Pick at least one space you have clean and organized at all times; revisiting here helps you center yourself.
- Add the color yellow. Wear yellow clothing or sit on a yellow cushion, cloth, or piece of paper while working at your desk. Combine yellow with the color turquoise to help you become clear about what you want while remaining true to yourself.
- Again, pack away extra projects; put them in boxes and behind closed doors. Be sure you have a completely clear workspace, clear desk top, and clear floor. Even clean out your email inbox and computer files.
- Check again you are in your power position. Be able to take two steps out in front of your desk. This space needs to be clear. Be sure you can swivel your chair all the way around in your work area.
- Remove any excess noise from your environment. Radio stations, TV shows, negative reading material. Be selective with what you accept into your mind space and be sure it is relevant to how you want to feel.
- Have an outside garden or some inside potted plants and spend a small time each day watering, weeding, and pruning

(connection with plant life will help ground you). Walk on the ground with your shoes off.

- Increase your water intake. Drink at least two liters per day.
- Activate the **north** (Career Path) sector of your office from the Bagua Map provided.

Candle Ritual for Taking Control:

Purchase a new **turquoise cylinder-shaped candle**. Write your Feeling Overwhelmed Affirmation on a yellow sheet of paper. Place a picture of water or water creatures, three $1 coins, a moon, a compass or a clock, an owl, or image of a pathway (you can include something metal or a water feature here) under the candle with your yellow sheet of paper. Include a white, silver, or gold piece of paper or object to enhance this process.

Each day light the candle and say your affirmation out loud. Then spend 45 minutes (or 3 x 15 minute sessions) visualizing what your life looks, sounds, feels, smells, and tastes like having achieved your desired 'in control' state. After this time put the candle flame out. Do this each day until the candle is finished. Journal your experiences and the changes in your *overwhelmed / control* state that occur. Perform this exercise in the **north** sector of your office.

9. Feeling like you have to do things all by yourself ...

Decide on an affirmation:

"Exactly the right people come to my business. I am encouraged and supported by all those I encounter along the way."

Ask yourself these questions:

- Does your office look empty or feel lonely? Are there pictures of single people, empty items? Does your workspace feel cold or dark?

Implement these Feng Shui strategies to address this challenge:

- Check for single photos and solitary items in your environment. Make use of symbols that represent support. A citrine crystal in your wallet or cash box can make you feel support in your abundance. Pictures of your mentors remind you of the assistance you do have. Images of angels signify that you are not alone and always supported by the Universe.
- Be sure you have a solid wall behind where you sit for work and meetings. Place a picture of a mountain or anything that represents support (elephants) behind you. Get an office chair with a high back and armrests.
- Identify and deal with any poison arrows that are intruding the space where you are sitting. These might be corners of furniture pointing at you, beams over your head, poles in the middle of a room, or open bookshelves.
- Add some warm colors to your environment like pinks, oranges, and yellows. Excessive use of cool colors (especially white) can make you feel lonely.
- Add movement to your environment. This can be live animals, plants, a mobile, a sliding photo frame, or a clock. Even images of live animals, plants, and people will work. Open a window for ventilation, even if just for a short period. Make sure your environment does not look too empty.
- Regulate heating / cooling so you are comfortable.

- Activate the **northwest** (Travel / Helpful People) sector of your office from the Bagua Map provided.

Candle Ritual for Implementing Help:

Purchase a **gold cylinder-shaped candle**. Write your Implementing Help Affirmation on an orange sheet of paper. Place a picture of a globe, your mentors or desired mentors, the stars and planets, a turtle, elephants with their trunks up, real coins or anything metal under the candle with your orange sheet of paper. You can even place a metal box filled with notes of what you need in your business beside your candle.

Each day light the candle and say your affirmation out loud. Then spend 45 minutes (or 3 x 15 minute sessions) visualizing what your life looks, sounds, feels, smells, and tastes like having attracted the helpful people you desire. After this time put the candle flame out. Do this each day until the candle is finished. Journal your experiences and the changes in your *help / assistance* state that occur. Perform this exercise in the **northwest** sector of your office.

10. Feeling like you are not good enough ...

"I know who I am and love what I do. My gifts are endless and I share them abundantly through my business."

Revisit the purpose exercise at the beginning of this book. Read your strengths and talents out loud and allow yourself to embody your giftedness. Then ask yourself this question ...

- Is your environment beautiful? Have you valued yourself enough to make this a special space for you?

Implement these Feng Shui strategies to address this challenge:

- Become conscious of anything that raises your energy and improves how you feel; add this to your daily environment. Commit to cleanliness, clarity, order, structure, beautiful surroundings, self-nurturing habits, flow, and ease in your office environment; reinforce that you value who you are and that you truly believe you are deserving of success.
- Have positive messages all around you. Display your awards and certificates. Put up photos of achievements and of people you have helped who are happy with your service. Place some testimonials in your work diary and on your website.
- Work with colors like pastel pink to encourage self-love or combinations like yellow and purple to enhance your sense of self-belief. Use green if you feel you need to heal something.
- Add a balance of feminine energy to your environment (as well as your attire) such as flowers, frills, hearts, soft and silky fabrics, feathers, and furs. Even men can do this in subtle and complementary ways; e.g., pink, purple, or yellow underwear, crystals in jewelry or pockets, floral art, a cushion added to a couch, a drape over a window, a piece of paper in a diary.
- Place a picture of you in the center of your Goal Board with all your desires surrounding you. Add the symbol of a heart (or the colors orange or fuchsia pink) to this space if you like.
- Activate the **northeast** (Business Belief) sector of your office from the Bagua Map provided. Have a collection of books or magazines that you love to read on display in this area.
- Buy fresh flowers for your office space weekly. You deserve to be acknowledged and feel special. The more you do this for yourself, the more others will do this for you as well.

Candle Ritual for Enhancing Business / Self-Belief:

Purchase a **yellow rectangular-shaped candle**. Write your Business / Self-Belief Affirmation on a purple sheet of paper. Place a picture of your most meaningful spiritual symbol, a mountain, an elephant, your favorite author, teacher or inventor, or a book you love under the candle with your purple sheet of paper.

Each day light the candle and say your affirmation out loud. Then spend 45 minutes (or 3 x 15 minute sessions) visualizing what your life looks, sounds, feels, smells, and tastes like having mastered the self-belief you desire. After this time put the candle flame out. Do this each day until the candle is finished. Journal your experiences and the changes in your *business / self-belief* state that occur. Perform this exercise in the **northeast** sector of your office.

The following story from a dear client shows the difference it can make when you believe you are worthy and you open yourself up to receiving wonderful gifts from the Universe.

I have to tell you an amazing story ...

You know how we were talking about flowers at the last Feng Shui lesson? I said that I never have any fresh flowers in my home – that I do not buy them for myself and there is no one in my life that gives them to me. Anyway, you asked me what type of flower I liked and I said 'gerberas.'

My daughter and I were on our way to the coast on Wednesday morning. I stopped in a town on the way to buy some fruit and a couple of salads for lunch. As I was waiting in the shop for the girl to finish the salads a man walked past, stopped in his tracks on the footpath, turned around and walked back into the shop. He had a silver beard and silver hair and was carrying something in his hand.

He then came up to me and said, "I thought this would go nicely with the dress you are wearing" and handed me a beautiful yellow gerbera done up in silver paper, which he had obviously bought from a florist!

I was dumbfounded and could only manage a smile and a "Thank you," along with a quizzical look. He left as quickly as he came and just disappeared around the corner.

I turned to the lady in the shop and said, "Is he a local and does he do this sort of thing all the time?" She said, "I thought he was your partner; I have never seen him before!"

My daughter was in the car across the road and said she saw what had happened. He had come out of the florist up the road and was walking down the street when he stopped halfway past the open door of the fruit shop, looked into the window, then turned around and went back into the shop where he handed me the flower. She said he just walked around the corner and disappeared!

I had only just been thinking to myself whilst driving that I was deserving of love and that I am ready to accept love into my life ... what was it that I was so afraid of anyway ... and that I could open my heart to warmth and love, plus I am ready to accept friendship and companionship.

Sophia

ACTION STEP #8: OVERCOME YOUR CURRENT CHALLENGES

Here is a list of questions to ask yourself to identify your current challenges:

Am I thrilled with the amount of money I make each week? Does it more than cover my bills, provide plenty to invest, save, give to my favorite causes, and use to expand my business? **Yes / No**

Do I effortlessly attract new clients who are eager to do business with me? Does this happen in regular cycles that make it stress-free for me to fulfill their needs? **Yes / No**

Do most of my clients stay with me over the long term? **Yes / No**

Do I know where I desire to take my business over the next six months, the next year? Do I know exactly what areas and aspects of my business to focus on? **Yes / No**

Do I feel that my time is expansive so that there is plenty of time to do all the things I desire in my business? **Yes / No**

Am I satisfied with the amount of time I can spend with family, friends, and doing things outside of work? **Yes / No**

Does my business get better every month (more exciting, expansive, fulfilling, and prosperous)? **Yes / No**

Do I feel a deep sense of mastery and power to create my business exactly the way I desire? **Yes / No**

Do I feel that I am surrounded by supportive partners, employees, and family members who collaborate in their own ideal way to co-create success for my business? **Yes / No**

Do I have a sense that I am worthy and that the work I do has value that is easily recognizable to everyone around me? **Yes / No**

Based on your answers, create three of your own positive affirmations to use daily in your business to address any areas where you are experiencing a challenge. You can use your own wording; just remember to make it positive and 'now' based. For example: "I effortlessly attract new clients every week who are eager to do business with me."

Alternately you can pick an affirmation from the Overcoming Challenges section above.

1. ______________________________

____

____

2. ______________________________

____

____

3. ______________________________

____

____

Now pick the affirmation that is most pending for you right now, create or adopt one of the above **Candle Rituals,** and begin a daily practice until the candle is finished or the challenge is overcome. Write down your experiences and the changes that happen over this time. *I will be so eager to hear your results.*

Chapter 9

HOW TO EXPRESS YOUR TRUE SELF-WORTH

Simulating Your Vibrational Pattern Associated With CONSTANT Financial Growth

Now that you have used Feng Shui on your business branding and your office space, let's circle back around to the topic of money again. Feng Shui can be used to improve all areas of your life, and your finances are always an important part of the big picture. Hopefully by this point you are starting to see why the energy represented by money is such an important aspect of your life. Wealth is simply a form of abundant, positive energy that overflows to everyone around you.

Money actually has a very powerful spiritual energy. That's because our culture invests it with so much value. In fact, the ease with which you attract wealth by doing work that you love is an excellent energy barometer. It demonstrates the amount of mastery you have achieved in directing the flow of Chi through your business.

Let's take a look at the Feng Shui of your finances and how you can enhance the good Chi in this area of your life.

FENG SHUI FOR FINANCES

First, you need to take a clear look at where you currently stand. If you aren't in the habit of doing this, it can be very unpleasant. This is especially true if you know you have quite a bit of debt spread around among multiple credit accounts.

However, when you don't value your money by tracking what is coming in and going out, you are sending a message to the Universe that says, "I don't have a good relationship with money. I avoid it." That needs to stop now. Reclaim the power to determine your financial situation by taking responsibility for it!

Here are just the first steps you can take to start seeing your current situation clearly:

- Balance your checkbooks for your personal and business bank accounts.
- Create a budget detailing your monthly household income and expenses.
- Create a second budget detailing your monthly business income and expenses.
- Make a simple Excel spreadsheet with all your assets (checking and savings accounts, retirement accounts, stock market investments, home equity, etc.).
- Add a second worksheet for all your debt (credit card debts, education loans, car payments, mortgage, business loans, etc.).

Include the interest rates and expected payoff dates of each one.

Based on what you discover during the process outlined above, take one action today to improve your money management practices. And remember; if you find all of this just too overwhelming then find the right finance professional to help you. It is crucial that you pay attention to this area of your life since it is such a big part of it.

Start Saving. It's Great Feng Shui!

If you don't have a savings account, now is the time to start one and begin 'paying yourself first' in monthly deposits, no matter how small. Why is this important? The Feng Shui vibration of a personal savings account says, "I value myself." Give the account a nickname such as The Wealth Builder or My Personal Prosperity Magnet. Don't call it your Rainy Day or Emergency Fund. That's like asking the Universe to send you an emergency that will wipe out your funds.

Don't have your savings account linked directly to your business account where an overdraft will draw down your savings. Your overdraft protection needs to consist of:

A. Knowing how much money is in your business account at all times because you value money enough to pay attention to its flow in and out of your life.

B. Making an agreement with yourself to always leave a minimum balance in place as a financial buffer.

Your online or mailed bank statement has a real vibrational impact on you (that's why so many people avoid looking at theirs). When you constantly let your balance hover right around zero that is some really negative Feng Shui you are creating in your outer environment. In contrast, when you look at your statement and see that there is a cushion of money in place, this generates an internal expectation / belief that you will always have money supporting you. Then, this is what you will continue creating.

Create Exponential Growth Potential

Keep on raising the amount you deposit in your savings account as the opportunity arises. Also, increase the amount of money you accumulate in your business account as a buffer. The vibrational pattern associated with constant growth is something the Universe picks up on and sends back to you; it's like a positive snowball effect.

This is one reason people with money always seem to have an easier time making more. It's not just because they have funds to reinvest in their business. They actually have an expectation that one of the **energetic purposes** of money is to generate (attract and manifest) more money. They may not use these terms to describe what they are doing, but you can see it happening in the way they manage their money and their business.

Alisse's Example: I found that Feng Shui made a huge difference in my financial situation in just a short time. For example, I struggled to purchase a new car in 2003. Then, three years later I bought another car through the same dealer. He made a special phone call to me the second time around to say, "What is the rags to riches story?" And, "What did you do differently?" I actually went from having a very small amount in my personal savings account the first time around to a substantial amount more the second time. And this was just my own personal financial improvement. The only thing I had done differently was that I had consciously 'changed the energy' of my business and my life. I actually worked fewer hours, changed my business product to something I preferred to work with, and had another child, going from three to four children. I found that building wealth was NOT about working myself to the bone and neglecting what was important in other areas of my life. It was about deliberately changing my attitude and my focus, and Feng Shui was the perfect tool to assist me in doing this.

10 HOT TIPS TO FENG SHUI YOUR WALLET!

1. Love your wallet. You reach for this item whenever you engage in a financial transaction. Since you want to associate positive feelings with everything having to do with money, your wallet needs to make you feel good whenever you see it.
2. Pick the perfect color for your wallet. Black and brown are usually not ideal. Instead, consider red (focus), green (abundance), or blue (financial responsibility).
3. Make sure your wallet is clean and in good condition with no worn or broken compartments / clasps, etc.
4. Clean your wallet out at the end of each day to remove and file away any receipts. Receipts are a focus on money you have already spent and promote the energy of money flowing out of your life.
5. Invest in a good quality wallet. Don't skimp here; your wallet is a symbol of how you value yourself and your money.
6. Choose a wallet that is easy to use. When you open it all the compartments need to be readily accessible so you don't have to keep turning it around to get to what you need. You want to experience flow when you are working with money.
7. Always have some money in your wallet. That way you won't reinforce a sense of lack by opening it up and seeing it empty.

You can even make some fake bills out of green or yellow paper (green for abundance or yellow for optimism).

8. To add the feeling of abundance and attract wealth, place a small citrine crystal or some Chinese coins in your wallet. A small mirror is another good addition for doubling your cash. Remember just one of these choices is ample; you do not want to create the feeling of being desperate or obstruct the easy flow of getting your money in and out.

9. Make sure your wallet is just the right size. If it is too small, it doesn't offer room for you to increase your abundance. If it is too large, it can be cumbersome and give you a negative feeling whenever you handle it. Also remember that if your wallet gets too full to close, it needs to be de-cluttered.

10. Place a picture of your current financial goal or a written goal card in your wallet. This can go in one of the clear plastic holders so you see it whenever you open it.

Alisse's Example: Buying a new wallet is one of my favorite ventures. My current wallet was beginning to wear, so I treated myself to a new *Guess* wallet in a gorgeous lime green color. I am a lover of *Guess* accessories, especially when they sparkle and shine. I picked the lime green for two reasons: because it looked fresh and fruitful and it represented 'new growth,' which associated nicely with how I wanted to feel about my money at that time. Also, it was warm in tone and one of my signature colors. Unfortunately, it was too small to fit all of my cards, etc., and I could not close it.

I went back to the store and found a purple *Jag* wallet, which is the color for enhancing belief. I was happy with this vibration and all of my cards, etc., fit easily. I kept this wallet for six months. But every time I went to pull it out of my handbag it got stuck. It was just a little too cumbersome and I immediately felt frustrated when this happened. The annoyance that I felt when dealing with money was not ideal. Fortunately, my sister bought me a new *Fossil* yellow wallet for my birthday. Would you believe it is just right? Yellow is for optimism and clarity. I love this wallet. It is good quality, sophisticated in its style, and reminds me of how special my sister is to me. It is a gift and that is the ideal way I want to feel when I deal with money.

It's time to have some fun buying yourself a new wallet!

FENG SHUI YOUR APPEARANCE

Using Feng Shui for your business attire is a fun and exciting way to boost your success. By wearing a particular outfit intentionally, you can anchor success into your clothing choices and create a positive mindset that attracts abundance. Besides paying attention to the colors in your business wardrobe, you need to also make a distinction between work clothes and non-work clothes. If you tend to wear the same clothing for work, fun, and relaxation, this may be an indication that you are not establishing clear boundaries between these different activities. In the same way, wearing specific jewelry while you are working and taking it off when you are done is an indication that you are able to keep a good work / life balance.

Your Personal Feng Shui Makeover

The information in this section is a wonderful tool for both business-women and businessmen. **Guys, don't skip this!** You will find fun ways to kick your personal style up a notch and start dressing for success. Plus, you can use the same information to help you pick the perfect accessory as a gift for the special women in your life.

Client Story: Once businessmen start learning Feng Shui tips, they can really get into it. One friend who runs a business with her husband told me the following story:

"My husband and I recently had our photo taken to use on our business card and website. He got really excited about shopping for the perfect tie that had our two main colors in it. Then, he had the idea to wear his gold tie chain with it since gold is our accent color. He made sure that the tie chain was visible in the photo so our business portrait would feature all our brand colors. He likes Feng Shui because it makes sense and has easy instructions to follow. Plus, he can instantly see and feel the difference when colors and design elements are carefully selected."

Let's take a look at your current business wardrobe.

One common mistake businesspeople make in this area is using **too much** black. If black is **not** worn strategically, it can literally suck the energy and personality out of your appearance. When you wear all black all of the time it is like you are erasing yourself. This doesn't mean you can't wear any black, just don't fall into the trap of wearing it because it is safe or easy.

You may think that you will **only** be taken seriously as a businessperson if you dress in somber colors. That's not true. Choosing the right styles and colors will raise your level of confidence and this is what your client detects the most.

(Women) Ditch the all black business suits and start expressing your mood and personality with your unique business style. (Men) It's time to spruce up your shirt and tie collections.

Powerful Colors for Business Attire

(Includes all shades and tones of each mentioned color)

Red is a color that demonstrates strength, power, and life force. Are you nervous when you have to speak in front of a group of people? Slip into a red suit, shirt, or blouse to give you courage. Use this color when you are setting goals or beginning a new project. It can also assist you in overcoming obstacles, staying focused, and being motivated.

Orange stands for creativity, joy, and celebration. Wear this color when you need increased flexibility in your business. Use it to help you let go of old habits and make changes. A burst of bright orange reminds you to celebrate your accomplishments and fully express your creativity.

Yellow is the color of clarity, thinking, and learning. Turn to yellow when you need to make a decision or be more positive about your business. For example, if you need to clearly understand how to reach your target market, you can use yellow to assist you. This color shines with positivity.

Green vibrates with growth, abundance, and balance. Are you ready to move forward with your business and infuse it with new ideas? Bring out your green clothing. This color also helps you improve your financial situation and brings a touch of extra luck just when you need it. If you are stressing out, add a touch of blue to reduce your anxiety.

Blue is an empowering color that resonates with good communication, flow, and trust. Wear it when you need to speak out about your true feelings. Use this color to help you let go, instead of trying to control everything. Blue instills a feeling of trust in your clients.

Turquoise Blue speaks to your authenticity, being true to yourself, and being able to communicate this in all you do. Draw on this color to help overcome emotional drainage you may be experiencing in your relationships with customers or business partners.

Purple is a fantastic color to use when you need to believe in yourself. Incorporate this color into your outfit to give you strength. Put on purple when you are about to speak to someone who may not be open to your ideas. This color can also help you connect with the purpose of your business.

Fuchsia represents community, Universal love, and the spirit of service. Wear this color when you are working with a team and when you want to send out a message of love.

Pair brown, **black**, gray, and white with other colors; avoid all **brown** all of the time as this can make you feel stuck. Remember, a lot of **white** and **gray** eventually feels lost and without direction.

Special Note: How to use black strategically. I have added this note because I personally love to add a portion of black to my business attire. I love the definition and strength of it and the innovative way it is used in fashion and design. It is also a highly used hue in the autumn and winter fashion months in many regions, and I am aware that many of you reading this book will likely reach for it at some point in your business life. For this reason, let's take a look at good Feng Shui tips for wearing black powerfully.

Used as an **accent** color, black can be very elegant and defining. Pair it with a good balance of other colored clothing (to add more vibrancy) and to counteract black's overly dominating energy. Add equal amounts of white if you want to achieve a harmonious yin-yang balance. Also add metallics via fabrics, metal jewelry and accessories (gold, silver, etc.). *Metal* will specifically enhance the luck of black's *water element* (Black and Metallic Gold in balance are considered an auspicious Feng Shui combination used consciously).

Avoid wearing **all** black, **all** of the time. It is too draining for you, your business, and your clients.

What About my Personal Color Palette?

Both men and women can benefit from getting a professional image makeover. This includes having your skincare routine evaluated. Clear, healthy skin attracts wealth in Feng Shui. Besides a balanced skincare regime, wearing the right colors next to your face evens out your skin tone and leads others to look you in the eye. This is very important in business. To make eye contact with someone is to connect and have them concentrate on what you are saying. This builds rapport and increases trust.

If you have already had your colors done by a professional, you know what shades and tints enhance your positive features. This does not mean you can't wear any of the other colors. It just means there is a particular intensity, clarity, and undertone that is more complementary to you. That's fine; picking a particular shade of purple or green that looks good on you will still let you tap into the energy of that color.

Also, if you want to wear something outside your normal range, just wear a neutral color at your neckline (one from your color palette). That way, a color that complements your skin tone will be directly next to your face. You can wear whatever color you want outside of that.

FENG SHUI YOUR ACCESSORIES

You can combine colors for an extra boost of energy. Accessories are the perfect area of your wardrobe to add touches of second or third colors. Scarves, purses, briefcases, shoes, socks, pocket squares, ties, key chains, and jewelry offer limitless opportunities for adding extra power to your outfit. Remember, the way you incorporate positive energy into your appearance changes how you feel on the inside and how your clients perceive you.

Energized Wardrobe Color Combinations

Red and Green: helps overcome procrastination.

Red and Purple: great for success in your career.

Red and Orange: activates your creative energy.

Orange and Purple: assists you in knowing your worth.

Orange and Pink: aids letting go of control.

Orange and White: enhances innovative planning.

Yellow and Red: perfect for knowing what's best to do.

Yellow and Pink: makes you smile.

Yellow and Purple: raises self-esteem.

Yellow and White: enables you to think clearly.

Yellow and Turquoise (add metal): inspires authenticity.

Green and Pink: attracts romance.

Green and Orange (add fire): great when you feel stuck.

Green and Gold (add fire): freedom.

Green and Turquoise: for following your heart's desires.

Blue and Green: reduces stress and disharmony.

Blue and White: speaks dedication.

Blue and Gold (metallic): invites wealth and riches.

Blue and Orange (add metal): deservedness.

Purple and Green: ideal for achieving your goals.

Purple and Pink: wonderful when you need inspiration.

Shoes, Shoes, Shoes

Shoes are an ideal place to start since they indicate the direction you are going with your business. Plus, they are fun to shop for. Now, you have another excuse to buy them; you have to Feng Shui your shoe collection!

Wearing a new pair of shoes is actually a great way to invite a new direction or a fresh start in your business. Women have an advantage over men in this area since there are endless options for color in a woman's shoe wardrobe. Don't just fit in with the crowd, stand out! That's what extraordinary people do.

Always choose good quality footwear that don't make your feet hurt; you don't want to hinder your ability to get where you are going. If you feel unstable in high heels, choose something with a lower heel to keep you feeling steady.

The shape of your shoes can tell a story just like the color does. An open toed shoe resonates with a sense of relaxation. Pointy toes are focused and action oriented. Wear these when you want to be assertive in a meeting. A rounded toe tells your clients that everything is balanced and in order in your business.

How you store your shoes is also of great importance, as these represent clear direction and forward movement. Be sure they are organized and easy to find in your closet. Keep them in pairs and sit them with the toes facing outwards (as opposed to facing the back wall). This represents that you are focused and 'ready to go' in the right direction.

Feng Shui Jewelry Tips

Wearing jewelry is one of the most wonderful ways both men and women can raise their vibration in business. Most jewelry has no practical purpose other than to make you feel good. This makes it a very powerful expression of who you are.

The material of the jewelry sends a message. Silver represents intuition and completion. Gold is for wealth and optimism. Each gemstone is a different crystal with its own vibration. For example, citrine (yellow quartz) can be used to raise abundance in your business; amethyst enhances strength and belief; while an emerald assists you to open your heart. Explore the color and qualities of precious or semi-precious gemstones to find a wealth of meaning you can bring into your life and work. If wearing crystal jewelry isn't your style, you can still access the energy of crystals by carrying one in your pocket or placing it on your keychain.

There are so many talented artisans online today. You can have literally any symbol you want incorporated into a custom piece of jewelry. Maybe you need a feeling of stability and a reminder to pace yourself. Use a pair of tortoise cufflinks. Do you want to improve your ability to listen to others? A pair of bird earrings might be the perfect answer. Perhaps a golden sunflower ring or tiepin would be just the thing to bring exciting opportunities within reach.

Remember, a jewelry piece that circles around but doesn't touch is a sign of things left undone or a tendency on your part not to finish what you start. Stay away from spiky accessories that give off an aggressive vibration.

Another way you can use Feng Shui with your jewelry is by paying attention to where on your body you place the item. Different parts of your body represent different things (as described in the chart which follows).

The hands are for giving and receiving, so a bracelet activates this concept. Earrings represent your ability to listen whereas a necklace points to your ability to communicate and speak your truth. Each finger also has a distinct meaning. The ring finger is representative of relationships while the center finger is good for business. (There is a more comprehensive description on page 296.)

With metal and gemstones it can be helpful to do cleansing exercises from time to time. Washing them, setting them on a window ledge in the light of the full moon, or burying them in soil overnight are a few ways to get a 'clean slate.' If you bury your jewelry pieces, put them in a potted plant container rather than directly in the ground. That way, you will be sure to find them again! Trust me they do like to move in the soil.

What Are You Carrying Around?

Your handbag is also very closely connected with money. Choose auspicious colors like green to attract more financial abundance. To express your confidence that prosperity is on its way, try red.

Carry coins and notes with you (even if you usually pay with plastic). These are powerful physical representations of wealth. You can also Feng Shui your business briefcase by choosing one that is made of luxurious material, or one with a vividly colored lining. Your business daybook, journal, or notepad can also be selected based on color.

Little Details Matter

One really quick makeover you can do is taking care of your fingernails. Your hands are a symbol of your connection with others. We give and receive with our hands, so spending some time pampering them is a wonderful way to care for ourselves. Men and women can both benefit from getting a manicure to make their hands look and feel great with buffed nails and smooth cuticles.

Do you enjoy wearing nail polish? Change the color of your nails to match your mood. Maybe try a bright pink to represent compassion and Universal love. Or, get white French tips to help you find structure and completion on an outstanding project. Don't forget to paint your toenails too. If you usually wear close-toed shoes you can get away with any color you like. Maybe you are looking to make a change in your life such as branching out into a new area with your business. Paint your toenails orange to bring the flexibility you need for this transformation.

Grooming With Color

Your makeup and hair can tie in with Feng Shui as well. Choose warm, golden highlights to surround yourself with wealth. Choose a red lipstick to give you extra energy or pink as a reminder to love yourself more each day. However, do remember to choose the hues that enhance your features, especially when working this close to your face.

A Strong Foundation

Don't forget to pick your undergarments with the same care as the rest of your wardrobe. Your underwear is the first thing you put on when you get dressed and it is closest to your body. Are you giving a presentation and need a boost of confidence? Slip on a pair of red panties that morning.

Do you work in an environment or a profession where uniforms or a strict dress code are the norm? No one but you will know what lingerie you are wearing, so it can be as adventurous as you like. Guys, don't forget that these tips apply to you as well. Slip on a pair of boxers or briefs in your favorite color.

Sleep on It

What about the pajamas you wear to bed? Your subconscious mind is very active while you sleep. You can use the night time to work toward specific goals. Say you go to bed with an unresolved problem in your business. Wear purple PJs to open up your insight. When you wake up in the morning, you may find that you have realized the perfect solution during the night. If you just need to rest after a long day, choose a combination of blue and green to help you relax and unwind.

Body Chart: Have some fun using your jewelry, makeup, clothing, and accessories on these body areas.

Head: Raises Intellectual Understanding

Neck: Strengthens Viewpoint

Heart: Encourages Openness

Waist: Enhances Courage

Hair: Promotes Identity

Throat: Improves Communication / Increases Flexibility

Solar Plexus: Increases Self-Value

Naval: Stimulates Intuition

The Right Side of the Body Represents the Physical Element. The Left Side of the Body Represents the Spiritual Element.

Ears: Improves Hearing Others

Nose: Emphasizes Wisdom

Mouth: Encourages Speaking Truth

Eyes: Enables Seeing Things Clearly

Cheeks: Enhances Enthusiasm

Fingernails: Promotes Glamour

Hands: Enhances Giving & Receiving
(Right Hand – Giving / Left Hand – Receiving)

Arms: Represents Reaching Out

Legs: Assists Moving Forward

Feet: Enable Going in Right Direction

Wrists: Denotes Handling the Situation

Ankles: Enhances Support in Direction

Toes: Promotes Balance

Fingers: Holding On / Letting Go

Fingers Right:	Fingers Left:
Thumb: Success	**Thumb:** Family
Index Finger: Power	**Index Finger:** Passion
Middle Finger: Business	**Middle Finger:** Mentorship
Ring Finger: Creation	**Ring Finger:** Innovation
Little Finger: Communication	**Little Finger:** Healing

GETTING DRESSED BY THE NUMBERS

Do you remember finding your personal energy number in Chapter 7? This number gives you insight into your ruling element and what you can do to enhance your success. You can use this information when choosing your business attire and accessories.

Energy Numbers 1 or 9: Your success direction element is Wood

You will be at your best in comfortable clothing that is simplistic and moveable. Choose to wear beautiful shoes, columnar-shaped styles, natural fabrics, and floral, striped, or rectangular patterned designs. Accessories made from wood or with flower, tree, and leaf symbols are lucky. Wear shades of green to enhance your success.

Energy Number 3: Your success direction element is Fire

You will shine in flashy clothing that grabs people's attention. Wear catchy eyeglass frames, enhancing eye makeup or your signature colors to highlight your eyes. Choose energetic design style, angular shapes, and triangular patterns. Accessories made from leathers and plastic materials are best, even animal symbols and prints. Wear red tones including oranges, pinks, and purples to enhance your success.

Energy Number 4: Your success direction element is Water

You'll enjoy wearing unique and creative clothing that is soft and flowing. Wear fluid or sheer fabrics, velvety textures, and asymmetrical, wavy, and artistic designs. Choose vintage items, beautiful accessories, long garments and necklaces, and scarves or neckties that are attractive. Wear shades of blue to enhance your success.

Energy Numbers 5, 2, or 8: Your success direction element is Earth

You will love timeless, traditional clothing and classic styles that are comfortable to wear, especially around your middle (or alternately that emphasize your center). Choose square-shaped designs and patterns, and accessories and jewelry with crystals in them. Draw attention to your hands with beautiful rings and bracelets, as well as eye-catching fingernail designs. Wear beiges, yellows, and oranges to enhance your success.

Energy Numbers 6 or 7: Your success direction element is Metal

You will be particular about your clothing, enjoying innovative and designer styles and luxurious fabrics. Choose form-fitting garments, rounded and arch-shaped designs, and elegant accessories. Clothing with circles and oval patterns (or no pattern at all), and polished metal jewelry pieces. Clear crystals and diamond jewelry are favorites. Wear white, pastels, and metallics to enhance your success.

Special Note: If you have made the decision that **you** are your business brand such as for a public figure, a singer, a consultant, an architect, a designer, etc., you might use your own personal success colors and elements when creating your brand design. This way you complement the energy of your business vision in alignment with your own personal energy.

Interestingly, most clients find that as their business becomes more aligned with their purpose, its elemental type is already closely congruent with their own personal energy. How amazing is that?

Unexpected Side Effects

When you start to Feng Shui for your personal appearance, perfect strangers are likely to come up to you and remark on how fantastic you look. Learn to accept compliments graciously without making self-deprecating remarks. Just smile and say, "Thank you, this color always makes me feel: confident, relaxed, creative, etc."

Change Starts Now

Make a commitment to go through your closet today. Get rid of any clothes that don't make you feel good about yourself. Then, you can start building a brand new wardrobe with pieces that bring out the best in you.

PAYING IT FORWARD WITH FENG SHUI

Using Feng Shui as outlined in this book is a huge gift that you are giving yourself. By now, you will be starting to feel pretty good. It's natural to start wondering what kind of gift to give to others as you begin overflowing with positive energy. Self-worth is one of the most wonderful gifts you can share with anyone. It **is** possible to use Feng Shui to assist you in this exercise. You are giving out of the abundance of the Universe flowing through you, and you will never be depleted when you give like this. This fact is in alignment with the Law of Cause and Effect. We are all meant to be deeply connected to one another and pass energy on in a cycle of giving and receiving.

Colors that will support you in sharing the vibration of Self-Worth:

Fuchsia Pink: enhances Universal love.

Purple: enhances belief in self.

Turquoise Blue: enhances being true to yourself.

Purple and **Orange:** enhances value of self.

Green and **Pink:** heals your relationship with yourself.

Feng Shui tips that will support you in sharing the vibration of Self-Worth:

Lotus Flower: connects with the soul.

Elephant: enhances strength and self-belief.

Cat: enhances self-esteem.

Rose Quartz Crystal: enhances self-love.

The following quote from the movie *Pay It Forward* draws a wonderful picture of how this kind of giving can work on a huge scale:

"If someone did you a favor, something big, something you could not do on your own, and instead of paying it back, you paid it forward to three people. And the next day they each paid it forward to three more, and the day after that, those 27 people each paid it forward to another three, and each day everyone in turn paid it forward to three more people, in two weeks that comes to 4,782,969 people."[14]

Note from Alisse: I remember watching the movie *Pay It Forward* and being deeply impacted by it. We can sometimes feel like we are too small to make any magnificent changes in the world and yet this movie showed us how our actions can significantly impact others. It also showed us that what we send out into the world comes back to us in unexpected times and ways. The question becomes not, "How can the world serve me?" But rather, "How can I be of service to the world?"

What I propose is that we 'pay it forward' with the gift of SELF-WORTH. I remember in my late teens thinking if I could put 'Self-Worth' into a bottle and sell it I would be a millionaire in a breath. While I cannot say I have bottled it as such, I do believe I have a few strategies to enhance it.

We might be of the belief that to 'pay it forward' you need to give something of material value, but don't forget that everything on the planet is energy in some form or another. Science has shown us this. So, be it seen or unseen, it is a gift of energy. What I am suggesting is that we forward some 'unseen' energy. The one thing I know about this sort of energy is that it is free-flowing, contagious, and rewarding.

So how do you forward SELF-WORTH to another person? It is as simple as paying a genuine compliment to every person you interact with (emphasizing their worth); from your customers and clients, to the person that delivers your mail, to the shop assistant that serves you in the grocery store. Try things like … "I love those shoes." "I like that color on you." "This is the best service I have had all day." "You were just the person I needed to chat with today," etc. This is a gift of positive energy and it will immediately raise the other person's vibration (including their ability to attract more wealth into their own business / life). Without them even knowing it, they will continue on their day and pass on some of that positivity you have shared.

If you are not comfortable giving compliments, you will get better with practice. It is truly a case of the more you do it the better you get at it! You can actually have fun thinking up innovative and wonderful things to say to people. It also has benefits over and above just making someone else feel good. It will turn your focus to searching for positives. It will train you to speak in more optimistic language. Plus, it will encourage you to think, "What can I do for others?" In return, you will immediately feel good for doing it and you will attract more positivity to yourself. Positive language will raise the feeling of value in the other person. It will also raise your own value of yourself, so the gift just keeps giving!

ACTION STEP #9: POWER YOUR BUSINESS IMAGE

Wardrobe Assessment:

As you go through your existing business wardrobe, ask yourself the following questions about each item:

1. Do I feel good when I wear this?
2. Does it fit me comfortably without feeling constricting?
3. Do I really like this style on me?
4. Does the fabric feel pleasant against my skin?
5. Is it in good condition and repair, or are there tears, holes, stains, or worn areas anywhere?
6. Would I wear this to an important business meeting?
7. Is the color one that looks good with my complexion?
8. Have I worn this item within the last six months? If not, why? (Hint: Don't save your best business clothes only for special occasions. Make your day special by dressing in your best.)

Be sure to check for what might be missing:

1. Do I have at least one really high-quality business suit?
2. What other key pieces are missing from my business wardrobe?
3. Do I have a good assortment of colored blouses or shirts?
4. Have I got enough pairs of smart, comfortable shoes?
5. What accessory would really add 'zing' to my usual look?

Write down what you discover from doing this assessment and make a plan. De-clutter what you don't like or wear in your wardrobe and replace it with those items you need. Remember to organize your shoes. Good Feng Shui in your personal wardrobe creates ease and flow at the beginning of your day. You will approach business feeling confident, energetic, and empowered.

Chapter 10
VIBRATIONAL IMPRINTS OF BUSINESS INTENTION

The Power of Faith – Will – and Imagination

Clear and specific goal-setting is one of the most vital lessons you can learn for creating and attracting wealth in your life. That's because a goal is not simply a plan for the future. It is a vibrational imprint of your **intention**. The more attention you pay to creating and nurturing your goals, the greater power they have in your life. This means you need to be very sure that what you set your sights on is what you truly want to invite into your life. Bear these words of wisdom in mind when you choose a goal:

"The steps of getting there are the qualities of being there"[15]

– Lazaris

In other words, the point of a goal is really about who you **be**-come on the journey. Consider the people you know who are rich in their business and in their lives. The qualities they demonstrate at the pinnacle of their accomplishments are the same ones they developed and continually expressed on their way to the top. Characteristics like Faith, Will, and Imagination have made them who they are. Seen from this perspective, the achievement of a goal itself (such as financial success) is almost a side effect of the process rather than an end point.

It can be very enlightening to view your goal-setting in this way. Instead of asking merely, "What do I want to achieve?" you need also ask, "Who do I want to **be**?" This will have a profound impact on the way you describe and visualize your goals. It also means that you will pay attention to ensure every action taken in achieving a goal is in alignment with the type of character you are building.

Special Note: Now is a very good time to refer back to your 'Definite Purpose Statement' and reacquaint yourself with the state of being that brings you most joy. Your goals will be most powerful when they are in resonance with you becoming more of the 'beingness' that you love the most.

Worthy Goals

When you are setting goals there are several reasons to aim high. First, your goals determine how you invest your time and energy. If you are spending the precious days of your life on a goal, shouldn't it be substantial? After all, you just might achieve it.

Second, your goal needs to be one that makes you stretch; otherwise it will not aid you in ***be-coming*** (remember it is all about the journey). You need to be really excited and a little scared, in a good way, when you think about your goal.

This doesn't mean you set a goal you believe deep down to be totally impossible at your current level of understanding. Remember, you will attract what you feel, no matter what you say with your words, and often this approach is just a way to sabotage achieving it (making it a fantasy). So, start with a goal that is a stretch but within what you believe to be the realm of possibility for now. As you achieve your 'starter' goals, the world of the 'possible' will expand before you. Then you can set more and more exciting goals as you grow in your ability to create what you desire. Learning to believe in your own incredible power is one of the steps of *getting there* that is also a quality of *being there*.

Finding Perspective

A worthy goal is one that takes effort, but not drudgery. When you are doing what you love and it matches your purpose in life, there is joy in pressing forward toward your goal. You may think you hate obstacles and fear failure. If so, it is time to obtain a different perspective. Replace the word 'obstacle' with the word '**opportunity**' and the word 'failure' with the word '**feedback**.' This is the reality you can choose to perceive every time a challenge presents itself on your path.

When you have clearly set your intention on a specific goal everything you encounter along the way is actually supporting you in achieving your goal. You just have to know how to identify it. Sometimes you don't realize until after a challenge is overcome, how much it helped you grow. For example, let's say you are a consultant and your goal is to increase your income without sacrificing the time you spend at home with your family.

You realize that you have been undervaluing your services and it is time to increase your pricing to match the true value of what you are offering clients. Then, out of the blue, your largest client (who accounts for a full 30% of your recurring business) informs you that they no longer require your services. There are two ways you could respond in this scenario.

You could become discouraged about your plan to increase your income and decide this is the worst possible time to raise your pricing. You might even decide to lower your pricing to see if you could get your old client back or rapidly find new ones who are looking for a bargain. You might even think you are being punished somehow for being 'greedy.' You could become obsessed with keeping your income from dropping over the short term and completely lose sight of your original goal. This means you work longer and longer hours at discounted rates just to keep your head above water. Spending time with your family takes a back seat, the exact opposite of what you had originally intended.

Or, you could view this as an opportunity to use the extra time you now have available to learn powerful ways to attract new clients who value your services more highly. As you build this client base that is paying your new rates, your income stabilizes and then begins to rise. Because you don't have to work extra hours just to keep food on the table, you can spend more time with your family. Furthermore, the hours you are investing in continuing to grow your business feel more fulfilling now. You are valuing your work appropriately and this makes you feel energized instead of worn out and resentful whenever you are in the office.

When you look back on this whole chain of events, you feel deep gratitude for the day your major client walked out on you. It becomes clear that this was a gift since it created room in your business for the growth and prosperity you deeply desire.

EFFECTIVE GOAL CREATION

Goal creation doesn't have to be a hit or miss proposition. There are a number of guidelines for setting and achieving goals. Here's a detailed breakdown of practical goal-setting. In this analysis, a goal is anything you desire to do, be, or have in your business and your finances.

Prioritization

It is great to have more than one goal at a time. However, it is better to prioritize them based on which are most meaningful and important to you. When you prioritize your goals, you are placing your focus and intention more specifically rather than having it remain scattered and nebulous. This sends a clearer message to the Universe about what you really want.

Phrasing

Just as for your affirmations it is absolutely essential to phrase your goals as if you already have them. As mentioned earlier, the subconscious mind cannot tell the difference between what is real and what is not; it accepts everything. This means the process of building a strong internal belief works best when you state things as if they 'are' rather than as if they 'will be.' Here's an example of a business goal stated in the now: "My new office is large and spacious with plenty of room for all the clients I am attracting. It is in the perfect location to continue boosting my business growth."

Specific + Measurable

Being specific about your goals is another important factor in success. For example, setting a goal to be richer than you are now is too vague. You might find a $10 bill lying on the ground later today. Hey, presto! You are richer. Not what you had in mind? You needed to be more specific with your request. As an example, a specific goal may be to increase your income by 25% and put $600 per month in your savings account.

It is also an excellent idea to be specific about *how* you will receive what you are asking the Universe for. This doesn't mean placing restrictions on abundance you will receive in your business. However, as Native

American shaman, Walks with Thunder, points out, it's a good idea to ask that gifts from the Universe come to you, "In a good and gentle way."[16] That way, if you need increased cash flow it won't come in the form of receiving an insurance check for your business burning down.

Time Limitations

There's nothing like a deadline to prompt you to take **action**, which just happens to be a critical aspect of goal creation. This means you need to put a date on your goal. It doesn't have to be exact or completely inflexible. It also doesn't mean that you are a failure if you don't realize a goal within the set timeframe. The date you put on your goal is simply there to remind you to engage in positive actions that promote your goal. Think about how much work you manage to get done if you know you will be going on vacation the next day. Deadlines for goals work the same way; they bring urgency and excitement to the process.

In essence ...

A Goal is a POWERFUL AFFIRMATION

written for your most highly Prioritized Business Desire

with a Specific and Measurable Outcome

required to Be Achieved in a Set Timeframe

Goals Must:

- Match your purpose (demonstrate who you want to be)
- Be clearly defined (so the Universe knows what you want)

- Stretch you but not break you (entice you outside your comfort zone)
- Be written in the now (to train your mind in a new reality)
- Be measurable (so you can track your progress)
- Have a date on them (so you are prompted to act)

Here is an example of this process:

Client Story: Kerry is a client who has always said she was going to create a website for her business, and write a weekly blog to educate her customers on choosing the right flowers for their wedding day. She owns a florist shop and finds her specific market repeatedly seeks out this same information.

While she has had this as one of her goals for a while she hasn't managed to find the time to organize it. Kerry has a deep love for providing the perfect flowers for her clients on their special day, her *business vision* is to be a recognized industry professional for doing this, therefore this goal does **match her purpose** and who she wants to be.

The first thing Kerry needed to do was to **clearly define** this goal. She wrote: "I want to create a website and write a weekly blog, educating brides-to-be how to choose the perfect flowers for their wedding day."

The next thing Kerry explored was: Did this goal **stretch her** (take her out of her comfort zone), and at the same time did she still believe it was possible? Kerry was a little afraid of taking her business online and quite reserved about writing a blog; she wondered if she was good enough to do this (writing was not her greatest strength), though at the same time she was very excited and knew that if she hired the right professionals and took it one step at a time she could successfully achieve this.

Kerry took my advice and wrote her goal **in the now** even though it was a new concept to her. It read like this: "I am so excited now I have created a website and am writing a weekly blog, educating brides-to-be how to choose the perfect flowers for their wedding day."

To make the goal more **measurable** she added some benchmarks that indicated Kerry had successfully achieved her goal. "I am so excited now I have launched my website and am publishing my weekly blog every Tuesday, educating brides-to-be how to choose the perfect flowers for their wedding day."

The final step of setting this goal was to **set a date** so that Kerry would prioritize this task and make time in her busy schedule to complete it. "It is the 1st of September 2015 and I am so excited now I have launched my website and am publishing my weekly blog every Tuesday, educating brides-to-be how to choose the perfect flowers for their wedding day."

Kerry felt this was a perfect date to launch her website as it was the first day of spring (where she lived) and the beginning of the wedding season.

GOAL SETTING FOR YOUR BUSINESS

Bringing it all together

Now that you understand the basics of goal-setting, here's a step-by-step action list to follow for creating / attracting / achieving / materializing your **BIG business goals**:

Step 1: Identify the business challenge / problem you want to change. Ideally, you state this in a single sentence.

Step 2: Flip the challenge by describing EXACTLY how you would like things to be in regards to this area of your business. Tell a new story.

Step 3: Turn the above description of how you want things to be into a goal following the guidelines you just learned. Use your goal as the affirmation that you read morning and night.

Step 4: Write this affirmation on a goal card to carry with you. This could be in your wallet or your journal, anything you regularly open. Read it (out loud if possible) whenever you put your hand on it. This will keep reinforcing the goal by bringing it into your conscious awareness and triggering the positive feelings associated with the goal in your subconscious.

Step 5: Include an image that represents this goal on your Goal Board. Place it somewhere you will see it a number of times each day.

Step 6: Feng Shui your environment to reflect your goal, e.g.: more wealth, an abundant environment, more clients, extra chairs in your waiting room, more recognition, ensuring you are the king / queen of your space.

Step 7: Visualize having achieved or attained this goal for a *minimum* of 15 minutes every day. (Note: The size of the goal and your level of belief in achieving it will determine the time you need to spend and how many times a day you visualize it.) Remember this is how it works its way into your vibration. Light a candle to assist this process.

Step 8: Experience the goal any way you can (as described earlier): attend the opening of your ideal office building, peruse your favorite successful websites, test drive your ultimate work vehicle, or try out your luxury office chair. This will assist your visualizing and make it more real to your mind.

Step 9: Remember and acknowledge that once your goal has been set it is truly on its way. These things simply take time; a natural gestation period if you will. Just trust and let go, knowing it is coming to you.

Step 10: Honor your destiny. When you write a goal, end it with the words, "That, or something better." This helps you to let go of the goal and allow the Universe to deliver it. Sometimes the Universe sends us a gift of what we need in a different package than we expected. When you accept this, you are honoring your inner knowing and your destiny.

Client Story: Kerry went ahead and followed these steps for the goal she created. It read like this:

"It is the 1st of September 2015 and I am so excited now I have launched my website and am publishing my weekly blog every Tuesday, educating brides-to-be how to choose the perfect flowers for their wedding day, that or something better."

Kerry did bring her goal into fruition only one week after the date she had set. She now posts her article every Tuesday as the goal suggests. Kerry has a growing following because of her blog posts and also thanks to her new website, local media has decided Kerry is the go-to girl for any information about wedding day flowers. She attributes a 40% increase in *new customers* thanks to all the opportunities that have developed from achieving this goal.

Further to this, Kerry has been approached by a publishing organization who suggested she turn her weekly blogs into a book filled with pictures of the beautiful bouquets she creates. This will give her the opportunity to present as even more of *the expert* in her field. She is now in the process of making this idea into a reality.

This was a good goal for Kerry to set.

YOUR GOAL BOARD. SEND A MESSAGE TO THE UNIVERSE!

One of the most exciting tools you can use in your office is a Goal Board. This is a collage of images and symbols that represent your goals. The best thing about this is the goals you choose can be literally **anything** you want.

Don't restrict yourself in what you imagine to be possible. If you think, "Oh, it would take me years to save up for a tour of Europe! I won't put that on my Goal Board," you are telling the Universe, "I don't believe I deserve to go traveling in Europe, so don't bother sending that opportunity my way."

Instead, be open to infinite abundance in your goal-setting. So, you wouldn't put up an index card with, "I will increase my client base by 5% over the next six months" written on it. Instead, put up a picture of a globe of the world (this stands for business expansion). Be patient and confident that the request you are sending out into the Universe is on its way to you.

You will be absolutely astonished at how many wonderful things materialize in your life and business once you physically put them up on a Goal Board. It is such a powerful way to reinforce your dreams by looking at these images every day. You are setting your intention clearly in your environment, which is the principle at the very heart of Feng Shui.

Put as many images as you want on your Goal Board. Remember, there isn't a limit. As you start achieving your goals and marking them off on your board, you will really start understanding how the power of Feng Shui and the Law of Attraction (Vibration) work together. The Universe has an endless supply of gifts; it is just waiting for you to ask.

The meaning of the images and symbols you use on this Goal Board are only part of the secret. The way you arrange them can add extra emphasis.

In Chapter 5 you learned how to use the Bagua Map for determining the layout of business cards, and it can be used the same way for making a Goal Board. Here's the map again for reference:

Business Wealth	**Business Fame**	**Business Relationships**
Business Health	**Chi Center**	**Business Creativity**
Business Belief	**Career Path**	**Travel/Helpful People**

As you build your Goal Board, you can create a collage of pictures, or just pick one image for each section, whichever approach feels most natural for you. The reason to use images and symbols rather than simply writing out a goal in each section is because pictures bypass the thought process and immediately impact the subconscious mind.

Tips: Golden Rules for Goal Board Pictures

- Don't use black and white pictures.

- For pictures of people, avoid those that have the top of their heads or legs cut off (waist and above or full length is best). Also, don't use pictures with people walking away or with their back to you.
- Avoid including images of people who have had a fateful, tragic life, e.g.: Marilyn Monroe, who died as a result of suicide. Characters in pictures need to look happy (if this is your goal!).
- The pictures need to look abundant. Avoid images of cactuses, long roads to nowhere, rough seas, stagnant water, deserts, etc. Just as you learned for all the images in your office space.
- Images need to be facing or moving towards the midline of the Goal Board. For example, if you are including a picture of your dream car, do not have it driving off the Goal Board. If you choose an airplane to represent travel, do not have it flying off the edge of the board.
- Watch out for waves and flowing water that are directed off the chart; this symbolizes money leaving your business.

Image Suggestions for a Bagua Business Goal Board

Following are some ideas for the types of images to use in each section of your Goal Board. For a Goal Board that is designed for your office, the more specifically the images represent your **business goals** the better.

For example, in the relationship corner you might put a picture of you and your team of partners to indicate that you are seeking powerful collaboration. Or, if one of your business prosperity goals is better work / life balance, you might put a picture of your family in that spot.

Business Abundance / Wealth Corner

Use pictures of waterfalls, water fountains, real money, gold bullions, bank checks (with your desired amount filled in), dream house, dream office space, sailing ships, water creatures, etc.

Also money plants (jade plants), frogs, leafy abundant plants, precious objects, and jewelry.

Enhancing Colors: *Green, Blue*

Business Recognition / Fame Square

Pictures of famous mentors and successful business entrepreneurs are great for this section. Include something that would have you know you are famous or that you have won; business awards, gold medals, trophies, articles of recognition, or a TV show / magazine you wish to appear in.

Also consider choosing a picture of the sun (not a sunset as this is used to finish something / let go of something), a phoenix, or birds flying.

Enhancing Colors: *Green, Red, Purple*

Business Relationships Corner

Use pictures of people interacting and looking happy, your business associates or clients, family pictures, animals communicating (especially birds), anything in pairs, a family home, a dining table with a bowl of oranges in the center, a romantic couple (if this is what you desire in your life).

Here you can include an image of rocks or crystals, even a tortoise.

Enhancing Colors: *Fuchsia, Yellow, Orange*

Business Creativity Square

Select pictures of people socializing, children playing, people having fun, images of famous inventions (e.g.: a light bulb), people using phones and technology, your own creations and pieces of art.

If you would like to birth a new business project (or start a family), pictures of pregnant moms, newborn babies, or an egg (golden if you like) will help here.

Enhancing Colors: *Yellow, Apricot, White*

Travel and Helpful People Corner

Select a picture of people shaking hands, an audience clapping, globes and world maps, foreign countries, holiday destinations, luxury vehicles, forms of travel, your dream car / boat / plane, planets, stars and space objects, friends, happy business employees and associates, your coaches, and mentors.

Enhancing Colors: *White, Silver, Gold*

Career Path Square

Use photos of water or water creatures (make sure any water is flowing toward you and not away), the moon, a pathway or road (not to nowhere), a desired pair of shoes, a key opening a door, a ladder, or your favorite form of travel.

Here you can place images of anything that will support you in your career; e.g., your title, business card, logo, money, future projects, etc.

Enhancing Colors: *Silver, Gold, Blue*

Business Belief Corner

Add pictures of your favorite reading books, powerful business success books, certificates, your teachers and mentors, and spiritual symbols. A mountain is a good image for support here and an elephant is good for belief.

Enhancing Colors: *Yellow, Purple, Ruby*

Business Health Square

Pick colorful flower images or baskets of fruit, your favorite meal, abundant trees, favorite ornaments, musical instruments, beautiful antique items, a feather pen and journal / diary, images of fit, healthy people or people at your ideal weight, any delicious healthy foods; also any symbols that represent health and longevity for you and your business.

Enhancing Colors: *Blue, Green*

Chi (Central Square)

The middle of the Chi area is a powerful location to place a picture of **you** (or you and your family / or you and your business team – dependent on the nature of the board). You can then write your Definite Purpose or words from your Business Vision underneath this picture if you like.

Enhancing Colors: *Beige, Earth tones, Rich Red .*

These suggestions are just a starting place; take your time to find the images that feel right and give you a positive boost every time you look at the board.

Alisse's Example: When you set your intentions in your outer environment with a Goal Board, get ready to pay attention when the Universe starts opening up doors for you to receive what you desire!

In 2008, I created a Goal Board that contained a picture of Wayne Dyer's latest book and *The Secret* DVD, which featured Bob Proctor. These two self-development experts are real heroes of mine and people I strive to emulate. I also placed a picture of an airplane (a symbol of travel) and a dragonfly (a symbol of transformation) on my board. I'd never particularly liked dragonflies before but felt drawn to them for some reason.

Within a few months of creating my Goal Board I received an email inviting me to become a business associate of none other than Bob Proctor himself!

Of course, this would entail flying from Australia to Florida. Handling all the arrangements would be quite challenging. As I was thinking through the implications of the invitation, I received another completely unexpected email. It was an update to let me know that there would be another mentor at the event with me. This special guest was Wayne Dyer. Up until that point, Proctor and Dyer had never worked together in the same room. Now they would both be at the same event and I would be working with two of my absolutely favorite mentors! This was truly a once in a lifetime opportunity for me.

However, because of all the obstacles I was focusing on (how much the trip would cost, if my kids would be OK while I was away, whether my husband could cope with running his business alone), I still hesitated. Every fiber of my being was telling me to 'just do it.' But I kept holding back.

Then, something really odd began to happen. I started seeing dragonflies everywhere I went. I asked other people if they had noticed a lot of dragonflies around and they always said, "No." I even found one sitting on the hood of my car. It was like this was a message just for me, saying it was time to take a transformative step forward. Still, because of my fears, I had not yet booked my flight across the ocean.

Then one day after I pulled my car into my garage and opened the door to swing my legs out, I saw a tiny live snake on the floor. The garage was sealed and I had not seen a snake in years, so I was shocked that it had gotten in. It was very small, but it still struck out at my leg fearlessly. Of course, I closed the car door and used my mobile phone to call for someone to come get that snake out of there!

As soon as I had the opportunity, I consulted my reference books about what the snake symbolizes. Just like the dragonfly, it is a sign of transformation. In addition, a snake striking is a message that a person is not 'striking out' and doing what needs to be done in their life. This was the final push I needed to book my flight to Florida and embark on what would be the most transformative experience of my life.

It doesn't really matter whether the Universe 'sent' me these signs or whether my subconscious mind simply engineered for me to notice them. I truly believe there are no coincidences and that we have the power to create every goal we set for ourselves.

Since that time, I have worked with Marie Diamond and Neale Donald Walsch, also from *The Secret*, and I have witnessed dozens of other pictures from my original Goal Board come to fruition (and I have made many more). I would never have believed how incredibly powerful this process actually is if I had not seen it for myself. I urge you not to take this information too lightly, and encourage you to make a Goal Board at your next possible chance.

It is time to initiate your own business magnetism with a new Goal Board for your office. Put it in a beautiful frame and place it in a special position using all the information you have learned so far. Or photograph it and get it printed on a canvas. Not only are Goal Boards amazing for business results they can also make a beautiful addition to your office décor.

THE 5 STAGES OF SUCCESS

An understanding of the five stages that you go through in achieving a goal can help you get a clearer idea of your progress. This concept is based on the five elements in nature and how they support each other. Without this ongoing cycle of support, nature would cease to prosper, and it works the same for business. If one element is too dominant it can stifle another, causing an interruption in productivity. Balance is key. We cannot ignore any of the phases if we hope to maintain continual success. That's tough to acknowledge sometimes, especially if there is one phase of business that we tend to like more than the others. For example, it may be easy for you to get excited and motivated about a new business idea but you feel less enthused about the phase where you are sustaining and growing your business over the long term. Understanding the cycle can help you appreciate the beauty and necessity of each phase.

Client Story: Josh is a client who owns a camping equipment store. His goal was similar to most business owners and that was to increase his profit. He is hugely passionate about his business, brilliant at buying stock that sells, serving customers, making sales, and managing the finances. However, Josh did not have a good marketing plan. This meant there was not a high enough volume of customers visiting the store. To improve sales, Josh would spend time sourcing better product lines and improving his customer service skills (because he loved to do this). While this would make small improvements in business, what Josh really needed to do was invest in a marketing campaign. Finally, Josh acknowledged this and hired a marketing team. His customer volume increased and sales doubled in the first month.

See how the missing element (Metal) was disrupting the Success Cycle and what happened when it was filled.

The list of the 'Stages of Success' below can be applied to any goal, short or long term, big or small. In fact, they apply to the operation of your **whole business**. When you are working toward multiple goals in your life or business (as most of us are), you might notice that you are at one stage with one goal and another stage with a different goal. This is a natural part of the growth and success cycle.

SUCCESS CYCLE

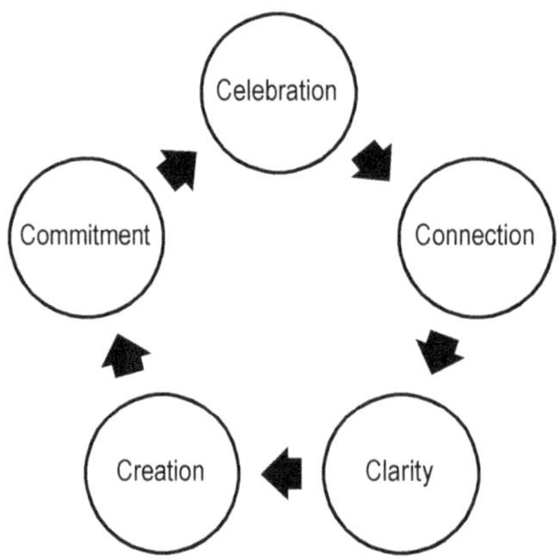

Here are the five Stages:

1. **Connection – Earth Phase**
2. **Clarity – Metal Phase**
3. **Creation – Water Phase**
4. **Commitment – Wood Phase**
5. **Celebration – Fire Phase**

Connection

Identify your life purpose and true desires for your business. Earth Energy = the foundation of your business. Fire (passion) fuels this process.

The groundwork must be done first, since this is the foundation of all that comes after. In business, it is common to misinterpret 'ground work' as being the creation of a good plan or system. That is actually the *next* step (clarity / metal phase). Systems and plans that are already proven to work may seem fine. But the catch is that they are proven *by someone else*. They are based on someone else's purpose and passion. They may work for a while. However, if you have not considered your 'Earth' (your truth), you will tire and feel unfulfilled. Most businesses fail because the person behind them has not worked through all the phases. You will find that goals are empty or unachievable if they do not have the right foundation for you.

In the connection phase, you create your Definite Purpose Statement. Connect with what will bring you joy. Decide what purpose or ideal is worthy of the investment of your time and then form your Business Vision.

Question: Do I know what my purpose is?

Feng Shui: In this phase you are becoming aware of your environment (conscious) and discovering what it is telling you about you and your business and your self. This occurs *before* you start changing things in your environment. Let it speak to you to show you clearly where you are right now.

Clarity

Understand what you will keep, what you will release, and how you intend to achieve your goal. This is the Metal Phase where you determine the architecture of your business. From the Earth, ore is taken to create something precious.

In this stage of clarity, you work with what you have learned about yourself from the Earth (connection) Phase. You get rid of what you don't need, keep what you do need, and create an action plan to get where you want to go / be. Once you have the point of focus (your Business Vision) you can build the path to your destination. At this stage, it is OK to find and incorporate great business, financial, marketing, and coaching plans that will take you on the quickest path to your goal. However, you have no need to chase shiny objects just because someone else says they are the key to success. You now know exactly where you are going and what you do and don't need.

Complete clarity in your plan is key; keep your focus on the plan and have a system to clear the path. Understand that limiting mental programming, clothed as fear and lack of belief, may become evident from time to time. Clarity requires you to have a plan to break through this (by identifying old beliefs and reprogramming your mind with new beliefs and intentions).

Question: What's the plan?

Feng Shui: Have you put clear plans and goals in writing? Place them somewhere you can see them so that you maintain focus. The plans and goals are all set in honor of your purpose, passion, and business vision. All decisions are made in reference to this information. In the office space, you maintain focus by first and foremost letting go. De-cluttering is a vital step. Get rid of everything that does not support your goal. Keep only what supports you in the best possible manner. Create a Goal Board and anything else that keeps your focus in the right direction, anything that reflects your goal in your environment.

Creation

Creation is the magic of your business. It is as much about flow as it is about work. This is the Water Phase.

Explore how you materialize your reality and learn tools to enhance your abilities in this area. Our need to control or feel separate and alone in the process of setting and achieving goals sometimes causes us to overthink them. This is why we so often focus on all the negatives of why what we desire is not possible. Then, this negativity cancels out our positive goals by sending a contradictory message to the Universe.

Remember this: You are the **co-creator** of your reality. While you are the planner, you are not the sole creator. This means you are not in complete control of the outcome, but it also means you are not alone. So, always plan ... but then *let go*. This is an extraordinary concept to grasp. Your goals are only guaranteed to materialize when you can trust and relinquish control. Accept that your best good is coming your way in accordance with Universal Laws and as part of a greater plan.

Question: Do I really believe this is possible? (Where is my faith?)

Feng Shui: This is where you learn to visualize, use candles, and send requests to co-create with the Universe. Trust in the power of consulting with your purpose and trust that everything has a gestation time. Relish the moment, let go and let God. Use Feng Shui antidotes to overcome challenges. Trust in your Goal Board. Trust that your environment is impacting your thinking. In business, you need a way to overcome stress, anxiety and being a control freak. Relax by taking six moments each day, to take six long deep breaths and just be present in your environment. Let go of control, allow success to come, accept that you are a co-creator and trust in the outcome.

Commitment

This is the Wood Phase, which is about the growth, health, and sustainability of business.

The commitment phase consists of the habits or actions you will do every day to nurture the growth of your business. You have set the goals / plan. Now, you need to put the plans into place and turn them into systems that you commit to for your success. What daily tasks and habits will you action? How will you nurture your love for your business? What is the action plan you will commit to?

Question: What actions am I taking (consistently)?

Feng Shui: Use this daily plan: Read your purpose and set intentions for the day, before your feet hit the ground. Check the Feng Shui of your workspace and ask if it reflects what you desire. Open windows, turn on adequate light, use a fragrant oil burner, choose your music ... create the feel. Write down your business vision and current goals in a journal and set three things you will do today which take you in the right direction to achieve your goal (you are about to learn a system for this in the final action step). Take your six breaths and visualize how you want your life to be. Follow your systems for filing, finances, marketing, etc. Empty rubbish and de-clutter your office space at the end of the day. Plus, keep learning, and growing, and investing back into your business for fresh energy and continuous new growth.

Celebration

Fire is about passion and joy. Celebrate your success and leverage it to bring even more passion and more joy into your life.

Remember that passion is the fuel, the Fire in your business. This keeps **it** alive, and keeps **you** fully alive. This phase is about setting benchmarks and really celebrating. Gift yourself time, kind words, material luxuries, and services that are a reward to you. This must be done daily in small ways. For example, take some time each day (even 10 minutes) to sit by the pool, walk in the park, read your favorite book, buy some flowers for your office, go and try on some new jewelry, or test drive your favorite dream car, etc. It is not OK to ever say you have no time to do this. That attitude and belief will just make your business *burn out*.

Some people never celebrate because as quickly as they achieve one milestone, they are looking for the next. There needs to be significant milestones of achievement that you celebrate as well as celebrating the small stuff. If you win an award, celebrate it. If you write another chapter for your book, celebrate it. If you manifest a parking spot close to the front of the shopping mall, celebrate it! Nothing is too small.

Question: What's the reward? (How will I celebrate this?)

Feng Shui: Hang up certificates, photos of significant events and achievements, places you have been recognized and featured. Always display the wonderful things clients say about you. Put these testimonials up in your office and on your website. Set regular benchmarks to celebrate and gift yourself with what brings you joy. This raises awareness and reinforces new behaviors. It also ensures that you will 'stay in love' and 'continue the romance' with your business vision.

The cycle continues as you go forward. The more Fire (passion, joy, love for your business) you have, the greater the value your service becomes to your clients. This raises your business energy and increases your business prosperity exponentially.

The 5 questions to check in with any business goal, venture, product, or endeavor.

1. **Do I know what my purpose is?**
2. **What's the plan?**

3. **Do I really believe this is possible? (Where is my faith?)**
4. **What actions will I / am I taking (consistently)?**
5. **What's the reward? (How will I celebrate this?)**

= The Productive Success Cycle

Client Story: Dana is a client who was able to make a huge business shift using this information. When I first saw her we used the 'stages of success' to see where her business was truly at and a real pattern emerged. While she felt confident in the Connection, Clarity, and Creation aspects, she noted she always started to peter out when it came to Commitment and Celebration. This resulted in what she called the 80% completion scenario. She would get so far in her ventures and then *stop growing* and *burn out*. Her habit was to conclude that she must be on the wrong track and she would simply change direction. As you can imagine, this was a real problem for her business success. She was not reaping the rewards and was feeling more and more frustrated. It was an eye-opener for her to identify what she was actually doing.

Dana had a real love for the researching, planning, and mastery of her belief patterns; however, she was not keen on the day-to-day actions. And when she did achieve something, she did not celebrate it as she felt there was still so much more to do. She realized it was not that she was on the *wrong track* but rather that she needed to master the commitment sector of her ventures, break projects down into daily tasks and continue with the smaller steps to achieve the bigger goal. She also realized that it was vitally important for her to embody her successes, no matter how small they were, and allow herself to feel the joy and accomplishment and reward herself. This was the actual key to her business success.

For some areas of her business she made the choice to hire other people, especially for the day-by-day mundane tasks she knew would deplete her energy. This meant they would still be attended to (she would still oversee them), however she could invest her time into her strengths and passions. She also realized that to truly value herself and her business she needed to invest the time into celebrating it. This was crucial in completing the Success Cycle and it was definitely a nice idea for Dana.

Dana is now completing her projects to 100% (sticking with the one thing until it is finished), which gives her even more motivation and enthusiasm to take on the next venture. With this, everything else has improved; her funds, her client volume, her opportunities, even her health. The best thing about these deep earthly insights is that they improve every aspect of your life, because you cannot raise the vibration in just one area without improving everything. Dana tells me she is in full flow now.

Using the five elements of the Success Cycle you can self-assess how much flow you have in your business ventures and project goals. If you can identify any weak spots, you will know this is where a little more focus needs to go. This way you can get your productive wheel spinning at full speed.

ACTION STEP #10: BEGIN YOUR SUCCESS JOURNEY NOW

Putting Your Goals Into Action

Now that you know how to materialize your goals, it is time to prioritize them. We will begin by turning your Business Vision into your Ultimate Goal. You will do that like this:

Alisse's Example:

Restate your Business Vision: To connect 1 million entrepreneurs worldwide with true wealth by having them create a business that expresses their infinite purpose and potential, so they can experience the freedom and fulfillment they desire.

Turn it into your Ultimate Goal: It is the 1st of January 2020 and I am so grateful now that I have connected 1 million entrepreneurs worldwide with true wealth by having them create a business that expresses their infinite purpose and potential, so they can experience the freedom and fulfillment they desire.

Restate your Business Vision:

____

____

____

____

Turn it into your Ultimate Goal:

It is the ____________________20____ and I am so grateful now that

____

____

____

____

Now write a list on a piece of paper of at least 20 business goals that will move you in the direction of your Ultimate Goal. (Yellow paper will help stimulate the process.) Once you have done this, highlight the 10 most important goals for you right now. A good way to do this is to ask questions like:

1. **Which of these goals is income-producing right now?**
2. **Which of these goals will make the biggest impact on my business growth?**
3. **Which of these goals will bring me the most joy while achieving it?**
4. **Which of these goals will take me most quickly towards my Ultimate Goal?**

Now take the highlighted goals and number them from 1-10 with the most important goal being #1. These will be your top priority goals for the current month.

My Top 10 Goals are:

1. ____________________
2. ____________________
3. ____________________
4. ____________________
5. ____________________
6. ____________________
7. ____________________
8. ____________________
9. ____________________
10. ____________________

Let's Get Journaling

Start a monthly journal. Pick a journal that is beautiful and abundant. Green is an excellent color for growth and wealth, purple is great for faith and expansion, and red for action, so all of these are good for a business goal journal. However, any color that you are drawn to strongly is fine (not black); it probably has a message for you about your business. Paste a picture on the front of the journal that represents a goal for your business. This might even be your business logo.

Open the journal and write your Ultimate Goal followed by your top 10 business goals for the current month on the first pages. The number 10 coincides with 'The Wheel of Fortune.' It invites action, as well as transformation and creativity. Each day, rewrite the 10 goals on the left-hand page of the journal. On the right-hand page write, "Actions I need to take today to achieve my goals."

List three tasks (and only three) that you **will** complete that day which will take you in the direction of achieving your goals. Then, *do what you committed* to for that day. Repeat this exercise every day. This is an awesome way to set your intention, boost your vibration, and enhance your materialized reality with positive action habits.

Alisse's Example: I think the most important thing when learning new information is to be able to integrate it into your business and your life. In this book you have learned a lot of tools, techniques, and strategies to make a big difference in your success. But like any new knowledge, it is only of value if you are able to make good practice of it. This is why I implement the journaling system.

I use my journal at the start and the finish of every day, and I include much of the information you have learned throughout this book in its pages. I record: my 'How I want my Business and Life to be' Story, Definite Purpose Statement, Business Vision, Ultimate Goal, Top 10 Goals, as well as my 3 Daily Tasks in it. Each day I rewrite my Top 10 Goals and decipher my 3 Daily Tasks. I have also created checklists of daily habits I chose to undertake (morning and afternoon) to ensure I am continually using Feng Shui for my MIND, ENVIRONMENT and SOUL (Heaven's, Human's, and Earth's Luck.) I start a new journal every 3 – 4 months, and I have my clients do this as well.

I realize this process is an investment of time (approximately 30 minutes twice a day) and that it does take a high level of commitment; however, my decision to do this is motivated by the words of

Bob Proctor when he said, "The line which separates winning from losing can be as fine as a razor's edge."[17] Successful people are the way they are because they go over and above what the average person will do. They go that extra mile and as a result they ensure greater rewards.

You can use the journaling system and design it any way you like; however, it is my recommendation that you do commit to this process. It is a way for you to become conscious, consistent, and accountable for your own business behaviors and actions, and to develop new and better habits (the secret of ease and sustainability). For me, I attribute the **fruition** of all of my successes (and those of my committed clients) to it. It really has been my greatest regulator for securing my dreams, goals, and desires.

CONCLUSION

As you continue on the exciting adventure of creating success, always remember that true wealth exists right here, right now! The way that leads beyond any feelings of defeat in business is to be **grateful** for all that is in this present moment. Shine the light on what you **do have** and **have achieved**; this is the ultimate 'cure' for everything you may encounter. Identify the wealth in your current life and in your existing environment, then nurture and grow it to match your infinite potential.

PRACTISE GRATITUDE

This is such an important thing to do every day. Always be grateful for what you have right now. It is only in this now moment that you can truly appreciate and enjoy all the gifts the Universe is showering upon you. Acknowledge your strengths and talents, the incredible business you do have, and all the amazing people and events that come your way each and every day (good and bad). Celebrate all the beauty that surrounds you, the potential available to you, and the stunning connection with a greater source of phenomena that wants nothing more for you than to reach your greatest version of success, right now. This is another necessary pathway to true health, wealth, and prosperity.

We are a society of people that are encumbered by scarcity and comparison mentality; as a result we hold on to the concept of *never enough*. We think we are: not tall enough, thin enough, fast enough, smart enough, wealthy enough. We tell ourselves that the obvious solution must be to have **more** of what we are seeking: more tall, more thin, more smart, more wealth.

The challenge with this concept is we continually find ourselves in the state of *wanting more.* We keep looking out into our future and back into our past not realizing that we only actually live in this now moment. What we perceive right now, by law, is our reality and we simply set ourselves up to attract more of what that is, *never enough* and *wanting more.*

What if we were to instead embrace that we have *enough* right now, and realized that this is the only actual solution to achieving more of what we desire? That only through mastering our gratitude (being able to see and appreciate what we do have) can we ever truly tap into our own ability to have the *enough* we are searching for. As counterintuitive as this might seem, it is an absolute necessary skill for attracting richness into our life.

Use your journal again and write down at least five things you are deeply grateful for at the end of every day. Speak these words of gratitude out loud to the Universe to invite more of these blessings to continue flowing into your life.

When times are tough or problems seem to arise, refer back to your gratitude lists and take a moment to focus on the abundance of prosperity that exists in your life right now.

Alisse's Example: At the end of every day I return to my journal and record two more things: 'What I did well today' (I recollect 5 things I believe I did well in that day) and 'What I am grateful for' (I list 5 things I am truly grateful for that day). I recite these out loud.

This is a very sacred process, the writing of my own personal journey. It is about my relationship with me and how I honor my business and my life.

Once I have completed these exercises I finish by lighting a candle and doing the 'Self-Love Celebration' you will read in a moment. I do this at my desk, snuff out the candle, close my journal, and acknowledge that my work for today is complete. This is the perfect way to finish my business day.

LOVE AND HONOR YOURSELF WITH ALL OF YOUR HEART

And finally there is one more thing we need to discuss. Something that aligns with your daily practice of gratitude. When you know that you are Divine, there is no doubt that your infinite ability will support you in achieving your goals. Speak this affirmation: "I honor the light and the life within me." No matter what is going on with your business, say this affirmation anyway. There is always Divine light and life within you regardless of what your outer circumstances appear to be.

Look into your own eyes in the mirror and simply say, "I love you." You are the one person who knows **YOU** completely and therefore the one person who can love you solely and unconditionally. When you love yourself you are forgiving your flaws and recognizing your perfection, and you will be enormously empowered to attract and receive abundance. Kim and Allison McMillen's book *When I Loved Myself Enough*[18] has many wonderfully empowering things to say about the topic of self-love. My mother gave me this book as a gift. Here are a couple of my favorite quotes from it:

When I loved myself enough:

- *I gave up the belief that life is hard*
- *I learned to meet my own needs and not call it selfish*

- *I realized I am never alone*
- *I began to taste freedom*

What a perfect example of how life can be when you learn to let go of the idea that you are not good enough. What will you do as you learn to love yourself enough? Spend some time thinking about all the ways your beliefs, your habits, and your life may change. As you learn to love yourself fully, there is no limit to the gifts you will be ready to receive in your life.

CELEBRATION OF YOUR OWN SELF-LOVE

Now it is your time to truly honor who you are.

Candle Ritual: Take a piece of green paper and a brand new pink egg-shaped candle. Write on the piece of paper: 'I love you, *Alisse* (substitute your own name there).' Write this five times. Fold and place the green paper under the candle. Enhance the exercise by including any of your favorite symbols. You may choose to use one of the Self-Worth symbols mentioned earlier (in Chapter 9), such as the cat or rose quartz crystal. (*Note: you only need to prepare this once for each new candle.*)

Then light the candle and speak these words: "I love you (your name). I truly honor and value you. You are worthy to me." Allow the candle to burn for a few moments while you reflect on all the wonderful things that you have, do, and are. Then breathe … 10 deep long free breaths of life.

As always repeat this each day until the candle is finished.

"I have come to realize that it is not talent or hard work, nor is it my ability to analyze and intellectualize, it is not a concept or a system I learn. It is the empowerment I have found by being in love with me. I have finally discovered the only answer to years of searching for fulfillment - it is - in fact - Self-Love."

- Alisse Bradley

Final Note from Alisse: As I look back over the course of my business and life it seems so surreal.

There was no practical reason for me to ever have been able to achieve the wealth, health, and success I have today. I didn't have a mass of money, education, or business smarts behind me. My belief in my abilities and myself was not abundant.

What did I do differently? Well, nothing in the **SEEN** component (or circumstances) of my business or life. However, in the **UNSEEN** (from within), I implemented Feng Shui and grew to develop an incredible understanding of the invisible force of energy that is the basis of everything we are, do, and have. I achieved everything I wanted from the *inside out* then the *outside in.* Some might say it was like a miracle but truthfully I embraced nothing more than pure and proven science.

I share all of my stories with you not to brag or impress you, but because I know you can have all of this and much more as well. Truthfully, if I can do any of what I have done, then most certainly you can too.

I deeply hope you have enjoyed reading this book and I thank you for it, but more so, I hope you fully embrace it and embody it into your own business and life. I hope you have learned that there **is enough** for all of us to have everything we ever wanted. That **you are** an infinite source of all potential, and can do anything you have ever dreamed of. And that everything you ever desire is **already within you** and **available for you** right now.

If it is your goal to double your income, earn six figures per annum, or build a million-dollar enterprise, you **CAN** (no matter where you are at right now), and you can do all of this and still live a valuable, meaningful, and purposeful life (in fact that's what you were meant to do). All you need to do is take what you have learned in this book and make it your daily practice, and promise yourself … *I will … until …*

I wish for you every success, every joy and a business and life that exceeds every expectation you could ever imagine. If I could only leave you with one thought, my gift to you, it would be this one.

You truly were born to live a RICH life (in every way)!

… so go and create, attract, earn, achieve and materialize the extraordinary business and life you were destined to live!

RECAP OF WHAT YOU HAVE LEARNED

The goal of this book is to give you actionable and transformational information for your business. Remember that the first important step is discovering your Life's Purpose; everything else flows from there. Here's a review of all the things you've discovered so far:

What you are meant to do in this life and in your work.

How your Life's Purpose translates directly into a Vision for your business.

1. The true nature and real value of the service you offer your clients.

2. How the Universe works and ways to align yourself with these natural laws.
3. How to tap into your Earth's luck and put it to work for you.
4. Ways to instantly attract your ideal clients.
5. Why you don't have to work 'hard' to be successful.
6. What your office space reveals about your current beliefs and habits.
7. How to effectively and accurately brand your business.
8. The meaning and uses of colors, shapes, images, symbols, and more.
9. How to raise the vibration of your marketing materials with intuitive design.
10. Ways to create the perfect, supportive office environment for you.
11. 10 ways to break through and invite success into your life / business.
12. Dozens of ways to raise the Chi of your business.
13. How to eliminate negative anchors and build new habits.
14. Fun ways to update your business wardrobe so it radiates positive energy.
15. How to set effective intentions and goals.
16. The importance of creating a Goal Board and guidelines for making yours.
17. The stages you can expect to go through as you set and achieve goals.

18. When and how to let go and trust that everything is perfectly designed to bring success straight to your doorstep.

As you begin implementing Feng Shui principles in your business, you will find yourself returning to this valuable resource again and again.

In the beginning, you may need to make sweeping changes in many areas of your business and office. Move at a pace that works for you and always take time to listen to your intuition. You will know that Feng Shui is working to enhance the flow of good Chi in your business when:

1. **You feel this energy on an emotional (vibrational) level**
2. **You begin to see the results in your business**

As you integrate these Feng Shui principles into your business, you will notice some changes right away. Improving the flow of Chi in your office space can brighten your mood instantly. Other changes are more gradual. For example, some people begin attracting more money immediately when they start implementing Feng Shui. For others, there is a longer period of 'gestation' between setting an intention and seeing the result. Just keep raising your vibration and taking positive actions toward your goals (remembering the mobile effect). The results are already on their way from the Universe to you. So, be prepared to receive wealth and abundance as it begins flowing (and overflowing) in your life.

I am so looking forward to hearing your inspiring stories of the business success you have achieved using the Millionaire Feng Shui methodology.

Share them with me anytime at: ***http://alissebradley.com/contact/***

Love, Light, and Laughter Always,

Alisse Bradley

ENDNOTES

1 Trump, *Donald.* "Donald Trump Quote." A-Z Quotes. AZQuotes, *n.d. Web. 09 Apr. 2016. <http://www.azquotes.com/quote/597635>.*

2 Chopra, Deepak. *The Seven Spiritual Laws of Success: A Practical Guide to the Fulfillment of Your Dreams*. San Rafael, CA: Amber-Allen Pub., 1994. Print.

3 Dyer, Wayne W. *The Shift: Taking Your Life from Ambition to Meaning*. Carlsbad, CA: Hay House, 2009. Print.

4 Nightingale, Earl. *The Strangest Secret*. United States: BN Pub., 2006. Print.

5 Hill, Napoleon. *Think and Grow Rich; the Famous Andrew Carnegie Formula for Money-making, Based upon the Thirteen Proven Steps to Riches*. Sydney: Angus & Robertson, 1959. Print.

6 Fondin, Michelle S. *The Wheel of Healing with Ayurveda: An Easy Guide to a Healthy Lifestyle*. Novato, California: New World Library, 2015. Print.

7 Diamond, Marie. *Diamond Feng Shui Essentials*. United States: Learning Strategies Corporation, 2006. Audio.

8 Grant, Eddy. *Walking on Sunshine*. Epic, 1978. CD.

9 *The Secret: The Secret Has Travelled Through Centuries To Reach You.* <http://www.thesecret.tv/products/the-secret-film-dvd/> Prime Time Productions, 2006. Movie.

10 *The Secret: The Secret Has Travelled Through Centuries To Reach You.* <http://www.thesecret.tv/products/the-secret-film-dvd/> Prime Time Productions, 2006. Movie.

11 Legge, James. *The I Ching.* New York: Dover Publications, 1963. Print.

12 Emoto, Masaru. *The Hidden Messages in Water*. Hillsboro, Oregon: Beyond Words Pub., 2004. Print.

13 Walsch, Neale Donald. *Conversations with God: An Uncommon Dialogue*. New York: G.P. Putnam's Sons, 1996. Print.

14 *Pay It Forward.* Warner Bros., 2000. Film.

15 <http://lazaris.com/publibrary/On_Deeper_Understanding_public.cfm>

16 <http://www.keen.com/CommunityServer/UserBlogPosts/Walks_With_Thunder/Meditaiton-for-Sept-15th/397658.aspx>

17 Proctor, Bob. *You Were Born Rich: Everyone is Born Rich in Potential, Most People Are Just a Little Short Of Money*. Scottsdale, AZ LifeSuccess Productions 2002. Study Program

18 McMillen, Kim, and Alison McMillen. *When I Loved Myself Enough*. New York: St. Martin's, 2001. Print.

FURTHER RESOURCES

Andrews, Ted. *Animal-speak: The Spiritual & Magical Powers of Creatures Great & Small*. St. Paul, MN, U.S.A.: Llewellyn Publications, 1993. Print.

Andrews, Ted. *How to Heal with Color*. St. Paul, MN, U.S.A.: Llewellyn Publications, 1992. Print.

Brazel, Christine. *54 Days to Fall in Love – "The Recipe": Feng Shui to Self Discovery*. Toowong, Qld.: Soul Energy 2003. Print

Brown, Simon. *Practical Feng Shui for Business*. London: Ward Lock, 1998. Print.

Chopra, Deepak. *The Seven Spiritual Laws of Success: A Pocketbook Guide to Fulfilling Your Dreams*. San Rafael, CA: Amber-Allen Publishing, Inc., 1994. Print.

Diamond, Marie. *Diamond Feng Shui: Balance, Harmony, Good Fortune*. Minnetonka, Minnesota, USA: Learning Strategies Corporation, 2006. Print.

Dyer, Wayne W. *Change Your Thoughts, Change Your Life: Living the Wisdom of the Tao*. Carlsbad, CA: Hay House, 2007. Print.

Dyer, Wayne W. *The Shift: Taking Your Life from Ambition to Meaning*. Carlsbad, CA: Hay House, 2009. Print.

Egan, Stephanie. *Colour Therapy Manual*. Mentone Aust.: Natural Energies College, 2002. Print.

Hay, Louise L. *Heal Your Body A-Z: The Mental Causes for Physical Illness and the Way to Overcome Them*. Carlsbad, CA: Hay House, 1998. Print.

Hay, Louise L. *You Can Heal Your Life*. Santa Monica, CA: Hay House, 1987. Print.

Hill, Napoleon. *Think and Grow Rich; the Famous Andrew Carnegie Formula for Money-making, Based upon the Thirteen Proven Steps to Riches*. Sydney: Angus & Robertson, 1959. Print.

King, Scott Alexander. *Animal Dreaming: The Spiritual and Symbolic Language of the Australasian Animals*. Glen Waverly, Vic.: Blue Angel Gallery, 2007. Print.

Lewin, Julie. *Areekeera Vibration: Healing Yourself From Within*. Milton, Aust: Celestial Consciousness Publishing, 2015. Print.

Linn, Denise. *Signposts: How to Interpret the Coincidences and Symbols in Your Life*. London: Rider, 1999. Print.

Noontil, Annette. *The Body Is the Barometer of the Soul so Be Your Own Doctor*. Nunawading, Vic.: Annette Noontil, 1988. Print.

Orman, Suze. *Women & Money: Owning the Power to Control your Destiny*. Alexandria NSW: Hay House Australia, Pty. Ltd., 2015. Print.

Proctor, Bob. *You Were Born Rich: Now You Can Discover and Develop Those Riches*. Cartersville, GA: LifeSuccess Productions, 1997. Print.

Shaffert, Eric. *Feng Shui and Money: A Nine-week Program for Creating Wealth Using Ancient Principles and Techniques*. New York: Allworth, 2002. Print.

Too, Lillian. *Lillian Too's Little Book of Feng Shui for the Office*. N.p.: Konsep Lagenda Sdn Bhd, 2008. Print.

Too, Lillian. *The Complete Illustrated Guide to Feng Shui*. Dorset: Element, 1996. Print.

Vitale, Joe. *The Attractor Factor: 5 Easy Steps for Creating Wealth (or Anything Else) from the inside out*. Hoboken, NJ: J. Wiley, 2005. Print.

Walsch, Neale Donald. *Conversations with God: An uncommon Dialogue*. Sydney, NSW Aust.: Hodder Australia, 1996. Print.

ACKNOWLEDGMENTS

My gratitude for those involved in the creation of *Millionaire Feng Shui: Discover the Hidden Science of Super Rich Business* is unbounded. I would first like to thank the funny little man at the spiritual festival (whose name I don't know) who read my palm many, many years ago and first planted the notion in my mind that one day I would write a book. How interesting these small little things are that get into the creases of your mind and sit there waiting for their turn to emerge. Until they do.

And from that point on there have been so many others.

My friend and mentor Danette Hibberd who offered me my first speaking gig on the topic of Feng Shui for the Millionaire Business – it all evolved from there. And to all those who interviewed me in the very early days, helping me to create my content and understand what entrepreneurs really want to know about Feng Shui, including Sue Papadoulis, Wendy Chamberlain, Lyndsey Baigent, Janet Beckers, and Pat Mussieux. You are all huge inspirations to me.

Notably to my Feng Shui Masters, teachers and trainers whom I have had the honor and privilege of learning this sacred wisdom from (some in person and others through their books and home studies). I would like to make special mention of: Marie Diamond of *The Secret* fame, whom I love and admire with all my heart; Chris Brazel, my first ever Feng Shui trainer and mentor who initially introduced me to the incredible power of this ancient science; and Nicole Bijlsma, founder of the Australian College of Environmental Studies, whom I continue to study with and learn from. I am so grateful for you all and have huge respect for the work you do.

Furthermore, to my book mentors past and present including Emily Gower, Kylee Ellis, Rachael Bermingham, and Robyn Henderson; because of the information you share I have gained the confidence and direction to publish this book for myself. And to my business mentors and wonderful friends Julie Lewin (my guiding light from the very first moment we met), Sally Thibault (my 'tapping' genius and step-by-step mentor), Deanna Cubis (my soul sister who was there for me every day), and Marguerite Louis Amor (who helped me clear the path and free myself), I thank you for all your inspiration and encouragement in finally getting this book across the line.

Also to Neale Donald Walsch, Bob Proctor, and the late Wayne Dyer who are shining lights in our world. True transformational leaders who have each shared the most encouraging and profound wisdom with me, assisting me to find within myself the belief that what I have to share is worthy. I will forever feel deeply blessed that I have had time with each of you in my life.

As for all those involved in getting this content into pure print the hugest thank you to you. Daisy McCarty and Assisi Chant, who guided me through the writing and editing of the book; Karen Thomson, who has been the marketing guru behind it all; and Bev Ryan Publishing for your incredible services in designing the cover, formatting the book, and getting it fully prepared for distribution.

Of course, I would be remiss in not thanking my gorgeous family: my husband Paul and four beautiful children; Anthony, Meagan, Renée, and Caleb. Writing a book and getting it to print has been a huge task for me so thank you for your patience and for holding the space for me to complete this project. It is a lifelong dream come true.

Finally, I would like to acknowledge myself. This truly has been one of the greatest personal challenges I have ever confronted and seeing it come into fruition is very rewarding to say the least. I can't say it was all smooth sailing or an easy journey; however, it has certainly been one worth taking.

And most especially to **you,** my readers; without your eagerness to know this information, your continual requests for me to put it into print, and the time you are now spending in reading it, there would be no reason for this book. For you all I am deeply grateful and I hope for you every success that this incredible ancient wisdom has to offer. I wrote this book especially for you.

ABOUT THE AUTHOR

Alisse Bradley is a unique Business Coach and Designer who combines 'energy based modalities' and 'Universal Principles' with savvy Western business techniques for fast and effective outcomes. She has devoted years of her time, energy, and love to studying disciplines ranging from natural, human, physical, and environmental science through to neuro-linguistic, metaphysic, and quantum science.

Beginning with a Graduate Diploma in Natural Science and an Associate Diploma in Human Science, Alisse's impressive and extensive list of qualifications includes several in human development including NLP, Time Line Therapy, Hypnosis, Diagnostic Astrology, Psycho-synthesis, Color Therapy, and Traditional + Western Feng Shui.

Overcoming a poverty-conscious young adulthood, Alisse knows what it means to live without wealth. Enduring a long journey to build self-esteem and liberate herself from believing that she was controlled by circumstances, Alisse overcame major challenges around money to live with the resounding certainty that she is her own best asset.

An accomplished speaker, magazine columnist, frequent radio guest, and the creator of her own training systems, she is sought after worldwide for her revolutionary and innovative coaching style and never miss results.

An amazing woman who fights for people's right to shine in this lifetime, she epitomizes the essence of wealth.

Deep belief in the transformation of the human being and the incredible workings of the mind-body-spirit lie at the core of what Alisse does so successfully and with such passion. A truly gifted speaker, Alisse communicates with ease, bringing a warm emotional connection to her audiences and clients. She delivers presentations, workshops, online programs, and one-on-one coaching / consulting to entrepreneurs, business owners, and solo professionals to show them how to make their lives shine and increase their financial wealth and personal prosperity.

She has trained personally with some of the world leaders in her field, including the acclaimed Bob Proctor of *The Secret* fame, the late Dr Wayne Dyer 'Father of Inspiration and Motivation,' Marie Diamond 'Feng Shui Expert to the Celebrities,' and Neale Donald Walsch 'renowned spiritual messenger' and author of the 'Conversations with God' series.

"It is my experience that money is the most spiritual thing on the planet," says Alisse. "Making more is, at its truth, a spiritual task; mastered in the mind and expressed on Earth."

"My mission is to guide entrepreneurs to live from the power of their purpose and the 'direct influence' of their energetic vibration. Within each and every one of us there is a burning desire to do something significant, pursue our greatness, and live through that extraordinary part of our selves. I want to raise people's conscious awareness and reveal their incredible potential so they can master their outcomes in life – including achieving true and sustainable wealth – and ultimately see them live the freedom and fulfillment they desire."

FENG SHUI AND BUSINESS

Would you like help to implement Feng Shui in your business and earn all the wealth, health, and success you desire?

YOU CAN WORK WITH ALISSE IN THESE WAYS:

- **Online Business and Wealth Creation Programs:** Alisse offers a number of online study programs you can do from the comfort of your own office to achieve your 'wealth and money' desires in your business and life.
- **Consulting:** You can consult one-on-one with Alisse about your different Feng Shui and Business needs.
- **6 month Feng Shui and Business Mentorship Program:** Join Alisse's membership club for six months and be guided to implement the Business Feng Shui principles outlined in this book.
- **Charts, eBooks and Audios:** Alisse offers a selection of charts, eBooks and audios for you to build your business / life consciousness for the purpose of increasing your business potential and profit.
- **Contract Alisse to work in your business:** Alisse will work personally in your business alongside you implementing the A – Z of Feng Shui for your business wealth, health, and success.

Go to www.AlisseBradley.com/Feng-Shui-Business to learn more!
Or visit www.AlisseBradley.com and to download
your FREE REPORT instantly!

FAST TRACK YOUR SUCCESS

How would your life BE if you could master your Wealth & Money state permanently?

THE 90 DAY 'WEALTH & MONEY' MASTERY PROGRAM

- Turn your **finances** around so that you never worry about money again.
- Discover your real **passion** and the **purpose** that lights up your days.
- Get rid of limiting thoughts blocking your money and allow it to **flow** to you.
- Find the **power** within you to change your life circumstances forever.
- Treat yourself to the life you **dream** of exactly as you want it to be.

This program includes:

Fourteen guided week-by-week lessons delivered online, exclusive access to The 'Wealth & Money' Mastery Blueprint, membership to the closed 'Wealth & Money' Mastery forum, as well as all the videos, templates, tools and resources you will need to fully claim your success.

Plus, you will learn the exact system Alisse used to master her own 'wealth and money' state, laid out in an easy step-by-step format, that anyone can follow and achieve, so that you can finally embrace the rich life you were meant to live.

Go to www.wealthmoneymastery.com and **START** today!